The Space Program
Quiz & Fact Book

THE SPACE PROGRAM QUIZ & FACT BOOK

TIMOTHY B. BENFORD
and BRIAN WILKES

Introduction by Frank Borman

HARPER & ROW, PUBLISHERS, New York

Cambridge, Philadelphia, San Francisco, London
Mexico City, São Paulo, Singapore, Sydney

1817

Photo credits: All photos are courtesy of the National Aeronautics and Space Administration.

Cover photo: Astronaut L. Gordon Cooper

FIRST EDITION

Layout: Timothy B. Benford

Library of Congress Cataloging in Publication Data

Benford, Timothy B.
  The space program quiz & fact book.

  Bibliography: p.
  Includes index.
  1. Astronautics—Miscellanea.   I. Wilkes, Brian.
II. Title.   III. Title: The space program quiz and fact book.
TL793.B395   1985   629.4   84-48816
ISBN 0-06-015454-3   85 86 87 88 89 RRD 10 9 8 7 6 5 4 3 2 1
ISBN 0-06-096005-1 (pbk.)   89 RRD 10 9 8 7 6 5 4

To the astronauts and cosmonauts, past and present, for
giving us mankind's greatest adventure. You do the things
we only dare to dream.

—Tim Benford

I had always wondered why authors traditionally dedicate
first books to their wives. After seeing what wives put
up with while authors struggle with first books, wives
clearly deserve much more. To Caryn.

—Brian Wilkes

Acknowledgments

"If I have seen further . . . it is by standing upon the shoulders of Giants."
—Sir Isaac Newton

Our deepest appreciation and sincerest thanks to the following people for their contributions:

Buzz Aldrin	Mike Collins	Wally Schirra
Neil Armstrong	Gordon Cooper	Deke Slayton
Frank Borman	John Glenn	Tom Stafford
Scott Carpenter	Fred Haise	James Webb
Eugene Cernan	Bruce McCandless	Al Worden

Our thanks also to the many current and former NASA employees at NASA Headquarters, Washington; Kennedy Space Center, Florida; Johnson Space Center, Texas; and Jet Propulsion Laboratories, California. Special thanks to:

Rocky Raab, NASA Public Affairs, KSC
Laszlo Dosa, Voice of America
Jerry Hannifin and Sue Butler Hannifin, *Time*
Ernie Legge, WMEL, Melbourne, Florida
Allen Moore, Esq.
James Edward Oberg
Bob Ward, *The Huntsville Times*
Kenneth M. Wing, M.D.

and, finally, to our families, from whom we stole the time for this project.

About the Authors

TIMOTHY B. BENFORD is the author of two other Harper & Row books, *The World War II Quiz & Fact Book* (1982) and *The World War II Quiz & Fact Book, Volume 2* (1984). His first novel, *Hitler's Daughter,* received a 1984 Porgie Award as one of the "best paperback originals of the year." A former newspaper and magazine editor, he is now president of his own public relations agency. As a freelance writer, his byline appears in publications throughout the country. He resides with his wife and two children in Mountainside, New Jersey.

BRIAN WILKES covered the Space Shuttle program and planetary exploration projects as news director of two Cape Canaveral-area radio stations and as science editor of an Orlando-based news network. He has covered the space program for CBS, Mutual, UPI, and AP radio networks and wire services, winning several awards for his work. He continues to freelance both broadcast and print media stories about space and the sciences. He now resides with his wife in Woodbridge, New Jersey.

The authors first met in a bookstore where Benford was appearing in connection with his first book. They soon discovered that *each* was then working on a book on the space program! The book that follows is the result of their joining forces.

Contents

Introduction

Mankind has looked to the heavens and dreamed of journeying to the stars since the dawn of recorded history and perhaps even earlier. Yet it was not until this century had moved beyond its halfway mark that the first exploratory steps to the cosmos were actually taken. I had the honor and good fortune to be part of this vanguard as a member of the second group of astronauts selected by NASA.

Authors and poets romanticized this quest even before it became a reality. Much will continue to be written as the frontiers of space are pushed even deeper and new goals and challenges are met.

But for now it is refreshing to reflect upon slightly more than twenty-five years of unequaled progress that certainly qualify for classification as the pioneer years of manned space flight. America mobilized its talents and husbanded its resources in what was touted as the Space Race with the Soviets. It was the largest single peacetime concentration of manpower and output from an industrial complex ever attempted. We accepted the challenge of sending men into space and returning them safely as first steps and then sending them to the Moon and beyond to fulfill the dreams of the ages. But all that was far away, it seemed, when NASA was still an infant agency in the family of government in the late 1950s.

This book does not pretend to be a scholarly examination of such history. Instead it is offered as exactly what it is: an interesting collection of little-known details, odd facts, anecdotes, vignettes, and superlatives served up in an entertaining and educational format.

It is a book to enjoy, to be sure. But perhaps even more so it is a book to remind us of what America can do when it accepts the challenge to be great.

—Col. Frank Borman, USAF (Ret.)
Gemini 7
Apollo 8

Q *Identify the trio of common denominators shared by the Original 7 Project Mercury Astronauts.*

A They were all first-born or only sons, all were white males, and all were Protestants. Seen here in 1959, front row, left to right: Walter M. Schirra, Jr., Donald K. Slayton, John H. Glenn, Jr., and Malcolm Scott Carpenter. Back row: Alan B. Shepard, Jr., Virgil I. "Gus" Grissom, and Leroy Gordon Cooper.

Manned Space Flights

Q *When was Project Mercury announced to the public, and what was significant about that date?*

A On December 17, 1958, the fifty-fifth anniversary of the Wright brothers' flight at Kitty Hawk, North Carolina.

Q *What is the significance of using the word "spacecraft" rather than "capsule" as it applies to manned space flight?*

A The Original 7 Astronauts were in agreement that "capsule" indicated that the man inside was only an experimental animal. "Spacecraft," however, described a vehicle that required a pilot to operate it. It was a change that they pressed in order to define their roles in the program.

Q *Who was the first man to land on a runway after flying in space?*

A If you said John Young or Bob Crippen, most space reporters would agree with you. But the correct answer is Bob White, who flew an X-15 to an altitude of 59 miles on July 17, 1962.

Q *What was the most unusual aspect in the background of the Group 7 astronauts?*

A They had all been astronauts before. The seven were Air Force officers assigned to the Manned Orbital Laboratory project. They were brought into NASA's astronaut program when MOL was canceled in August 1969.

Q *Identify the designer of the Mercury spacecraft.*

A NASA engineer Maxime Faget. Some early suggestions included spherical and cylindrical designs, and a winged craft. The shape Faget came up with has been compared to a gigantic TV tube.

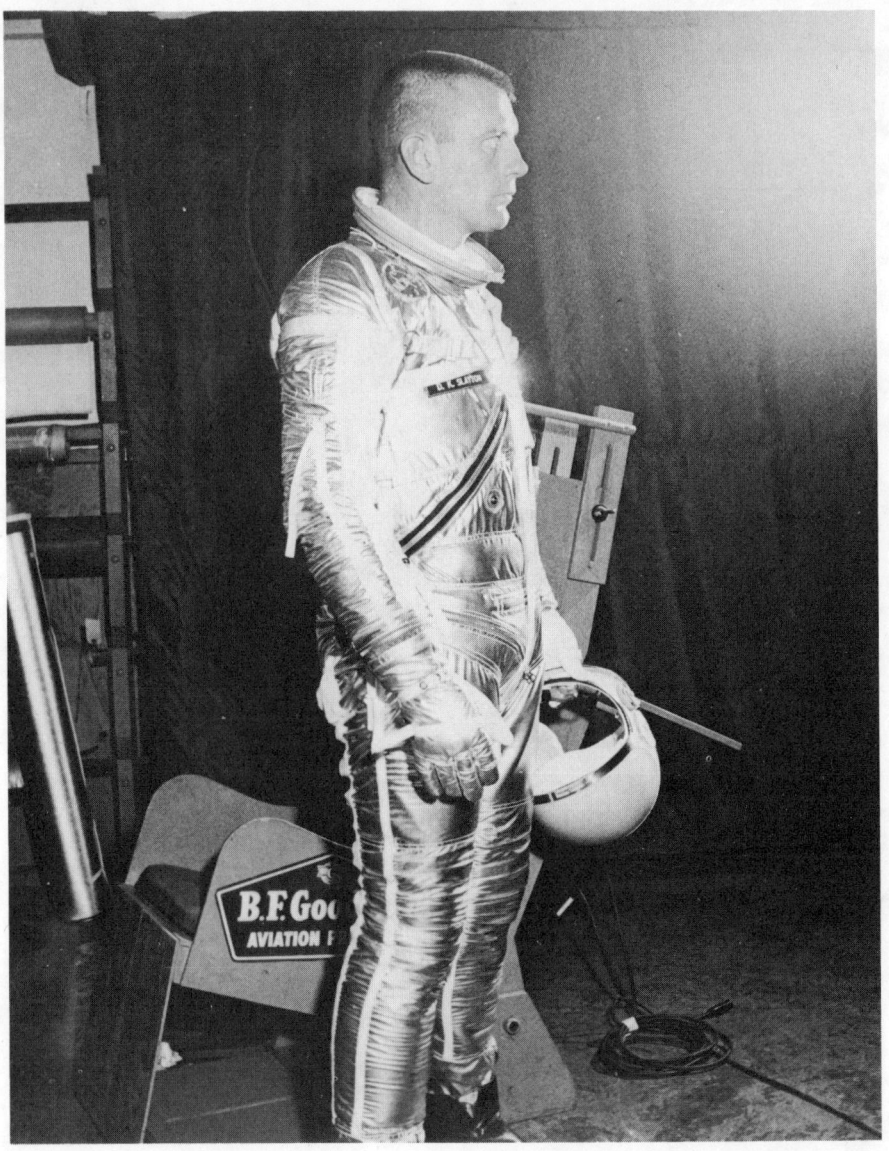

Q *What was the medical reason for Astronaut Deke Slayton's being grounded and not flying in Project Mercury?*

A He was diagnosed as having idiopathic atrial fibrillation, a mild irregularity of the heartbeat. It was first noted in August 1959 while he was functioning on the centrifuge. In this 1960 photo, Slayton is seen at the B. F. Goodrich plant, trying on a prototype of the suits the other six would eventually wear.

> **Small World Department** During an interview with Deke Slayton, Italian journalist Oriana Fallaci discovered that Slayton had been a pilot in the October 1943 air raid that destroyed her home in Florence.

Q *Identify the member of the original Soviet cosmonaut team who was grounded because of a "mild heart irregularity" uncovered during a centrifuge test.*

A Anatoly Maslennikov. The Soviet news reports failed to mention that the "irregularity" was discovered after a centrifuge malfunction had in fact overstressed Maslennikov and ruptured most of his internal organs.

Q *Who was selected to be the second astronaut to orbit the Earth?*

A Deke Slayton was scheduled to follow John Glenn, the first American in orbit, but a heart irregularity caused him to be grounded. Scott Carpenter replaced him.

Q *What medical problem bumped Mike Collins from the Apollo 8 mission?*

A While NASA releases call it merely a "nerve pinch" in his arm, it was much more serious. Vertebral pressure on the spinal cord, possibly caused by an ejection seat bailout, threatened the use of his legs. In a dangerous operation, surgeons went in through his throat and fused two of the vertebrae together. Although a simpler procedure was available, it would have left him grounded.

Q *Name the astronaut injured in the crash of a Japanese fighter plane.*

A Fred Haise, who suffered severe burns over most of his body in August 1973. He was ferrying a Vultee-Vibrator that had been made to resemble a Japanese "Val" dive bomber for the film *Tora Tora Tora*. The crash occurred at Harlingen, Texas, the home of the Confederate Air Force, a collection of antique planes.

Q *Who were the two Mercury astronauts who fought their way back to flight status after medical groundings?*

A Al Shepard and Deke Slayton. Shepard was grounded with an inner-ear defect early into Project Gemini; he flew on Apollo 14 after medical treatment (1971). Slayton was bumped from Mercury after a heart murmur was discovered. He was returned to flight status in 1972 and flew on the Apollo-Soyuz mission (1975).

Q *For what was Roger Chaffee best known before becoming an astronaut?*
A He took the high-resolution U-2 photo that revealed Soviet missiles in Cuba in 1962. These were the same pictures Ambassador Adlai Stevenson displayed at the United Nations. In this 1964 photo, Chaffee himself is in tropical climes, Panama to be exact, during astronaut survival training.

> **"Roger, Edwards, Déjà Vu ..."** During the shuttle *Enterprise*'s approach and landing tests, astronaut Joe Engle was dropped from a 747 piloted by Col. Fitzhugh Fulton. Must have seemed familiar, because years before, Fulton had piloted the B-29 that dropped Engle in X-15 tests, over that same lake bed.

Q *Identify the only one of the Original 7 Astronauts to have been a test pilot at Edwards Air Force Base.*

A Deke Slayton. Edwards was considered the ultimate in flight testing.

Q *Which astronaut was dropped from a mission because of a childhood disease he didn't have?*

A Ken Mattingly, who was exposed to Charlie Duke, who had German measles, before launch. Records showed Ken had never had measles, so he was replaced by Jack Swigert two days before the mission. Ken never developed measles, but he was tagged "Measles Mattingly" by some detractors in NASA. The incident led to a three-week prelaunch quarantine for all remaining missions.

Q *Which astronaut was removed from flight status following a motorbike accident?*

A Scott Carpenter, who had the crash in Bermuda in 1964. He never flew in space again.

Q *Which astronaut was injured when he slipped in the bathroom?*

A John Glenn. Although it was widely reported that he had broken his neck falling in the shower, he had actually been replacing a sliding glass shower door. He didn't fracture his neck, but he did suffer severe concussion. And the accident, which occurred on February 26, 1964, did more serious damage: it took him out of the Ohio senatorial primary.

Q *Identify the two astronauts who piloted the F-106 jets designated as chase planes during Alan Shepard's Mercury flight.*

A Wally Schirra and Scott Carpenter.

Q *Who worked as the capsule communicator (CapCom) during Alan Shepard's Mercury flight?*

A Fellow astronaut Deke Slayton.

Charm School Graduates The U.S. Air Force considered itself the only organization predestined to produce astronauts for NASA. It coveted this role so much that it created a "charm school" for its leading candidates to teach them how to act, dress, and talk in the company of NASA officials. Candidates learned, for instance, that they should wear knee-length socks so that if they crossed their legs while they were seated, no bare flesh would show. They were shown the difference between the ways a fighter jock and an interior decorator would stand with their hands on their hips (thumbs to the rear for jocks!). At social functions with NASA bigwigs it was permissible to drink bourbon or Scotch (no fancy lady drinks, please), but only one drink. It is interesting to note that of the seventy-two astronauts selected by NASA through June 1981, twenty came from the ranks of the Air Force. The rest were Navy, Marine, and civilian.

Q *Name the date and location in Washington, D.C., where the Original 7 Mercury Astronauts were presented to the public for the first time.*

A Dolly Madison House, on the northeast corner of Lafayette Square, at 2 p.m. on April 9, 1959. They are seen here less formally dressed while going through survival training at Stead Air Force Base, Nevada. From the left: Gordon Cooper, Scott Carpenter, John Glenn, Alan Shepard, Gus Grissom, Wally Schirra, Deke Slayton.

The Process of Elimination   When NASA began considering volunteers to become astronauts, it turned up a group of 508 men who met the basic requirements. This group was reduced to 110, then 69, and finally 32 who were put through exhaustive medical and psychological tests. Fourteen of them dropped out, and the selection of the Original 7 Mercury Astronauts came from the final field of 18. The initial field for the astronauts who became Group 2, or "the next 9," was 253 applicants.

Q *What were the selection criteria for the Original 7 Mercury astronauts?*

A Each had to be an active-duty military jet test pilot with at least 1,500 hours flying time, less than 40 years old, not over 5 feet 11 inches tall, and hold a B.S. or equivalent degree.

Q *Name the hometowns of the Original 7.*

A East Derry, New Hampshire (Shepard); Mitchell, Indiana (Grissom); Cambridge, Ohio (Glenn); Boulder, Colorado (Carpenter); Sparta, Wisconsin (Slayton); Oradell, New Jersey (Schirra); Shawnee, Oklahoma (Cooper).

Q *Who is the nominal head of the American space program?*

A The chairman of the National Space Council, who is always the vice president of the United States. This means that Walter Mondale, the most outspoken critic of the space program next to William Proxmire, was actually in charge at one time. Day-to-day affairs are bossed by the administrator of NASA. Other members of the Council are the secretary of state and the secretary of defense.

Q *Who is the nominal head of the Soviet space program?*

A The chairman of the Goskommissiya, or the State Commission for the Planning and Execution of Spaceflight. As with anything else in the maze of Soviet bureaucracy, it's almost impossible to tell who's really in charge at any given moment.

Coming in on Two Wings and a Prayer   When John Glenn splashed down in the Atlantic after his 1962 flight, he had two Wings to thank. Capt. Kenneth M. Wing was an Army doctor waiting aboard the USS *Stribling* to examine Glenn. His father, Kenneth A. Wing, was an electrical and aerospace engineer who had designed part of the Mercury spacecraft. Because of tight security, neither Wing knew the other was involved in the space program until months later.

Q *Name the Soviet spy who later became the head of the Russian space program.*

A Mikhail Yangel. A Soviet of German ancestry, he successfully infiltrated von Braun's rocket base at Peenemünde. After the war, he became Sergei Korolyov's assistant, taking over after Korolyov's death in 1966. He died in 1971.

Q *Who were the first spies in space?*

A Yuri Artyukhin and Pavel Popovich, who flew what appeared to be a wholly military mission aboard Salyut 4, starting July 3, 1974.

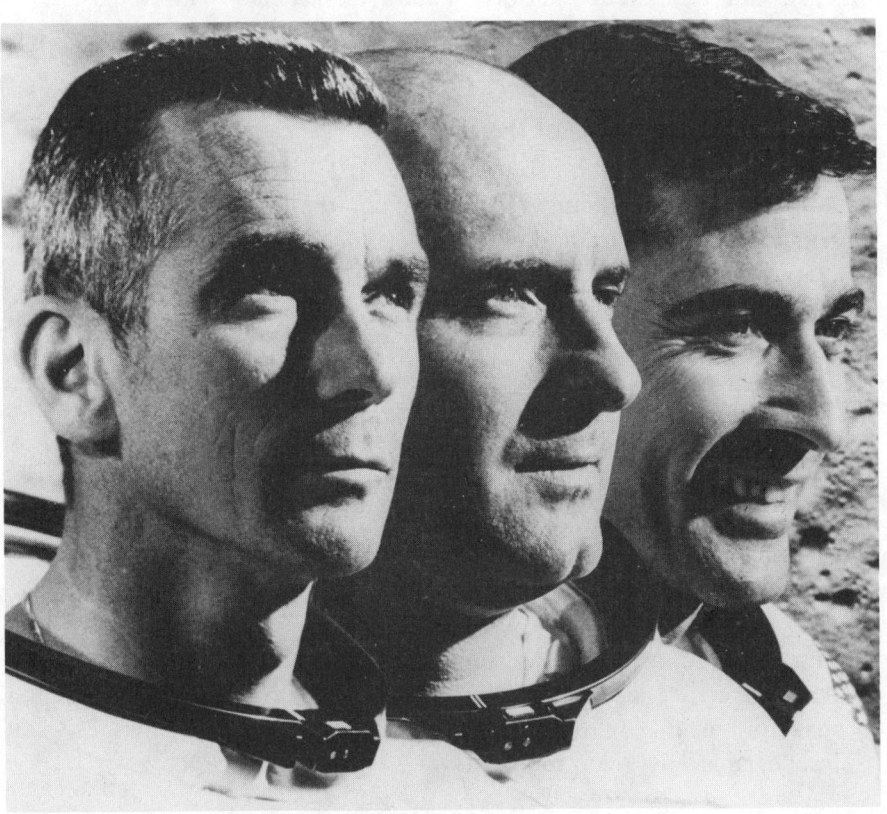

Q *Name the first astronauts to shave in space.*

A The crew of Apollo 10. NASA had previously feared that small hairs floating in the weightlessness of space might clog the delicate instruments or impair the astronauts' breathing. The preflight clean-shaven faces of Gene Cernan, Tom Stafford, and John Young are seen in this Apollo 10 crew montage.

> **Some Alternate!** John Glenn was not a first choice in the selection of the Original 7 Astronauts. He was assigned as an alternate when one of the other candidates failed one of the physical tests.

Q *Who got the first haircut in space?*

A Al Bean, on Skylab 3. The trim was given to him by Owen Garriott, while Bean used a suction hose to catch stray, weightless clippings.

Q *What unusual piece of medical equipment did Dr. William Douglas carry with him for Gus Grissom's flight?*

A A snakebite kit. Douglas was prepared to reach the astronaut with medical aid in the event the mission was aborted after lift-off. The area around the launchpad had a rather large rattlesnake population.

Q *Identify the first astronaut to wear a urine collection device.*

A Gus Grissom, on *Liberty Bell 7*.

Q *Who named John Glenn's spacecraft* Friendship 7?

A Glenn and his son and daughter, Dave and Lyn, made up a list of possible names. *Friendship 7* was selected by the youngsters. However, the spacecraft was officially designated on all lists as Capsule 13, since it was the thirteenth Mercury capsule to come to NASA from the manufacturer.

Q *Who painted the mission logo for* Friendship 7 *on John Glenn's spacecraft?*

A Artist Cecilia Bibby.

Q *Of the six Project Mercury manned flights, only two were recovered in the Pacific. Which were they?*

A Wally Schirra's *Sigma 7* and Gordon Cooper's *Faith 7*. The other four were recovered in the Atlantic.

Q *Which astronaut spotted a tugboat and barge from orbit?*

A During his Mercury mission, Gordon Cooper reported what he thought was an aircraft carrier in the St. John's River near Jacksonville. It checked out as a tugboat and barge. Cooper also described seeing highways and railroads, but those reports weren't taken seriously until confirmed by Gemini 4.

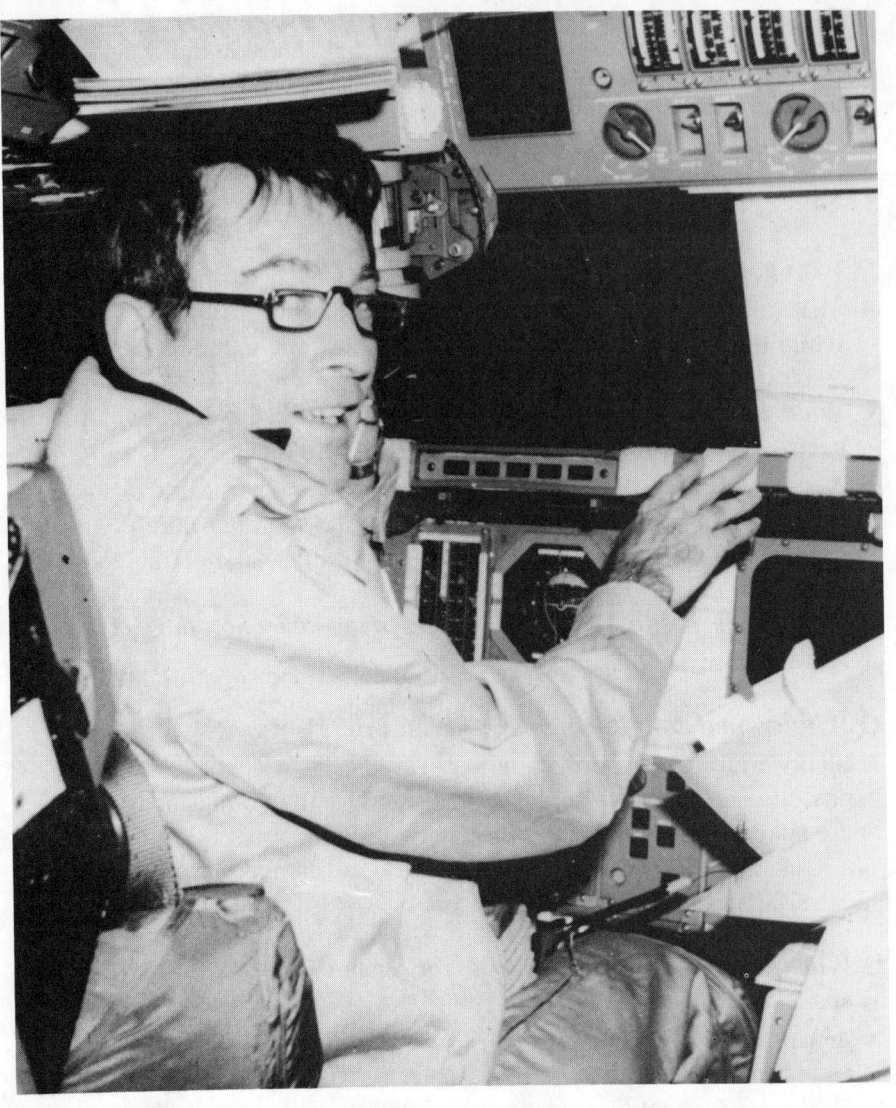

Easy, John, Easy—During the launch of the first shuttle (STS-1), astronaut Robert Crippen's heart rate rose to 135 beats per minute, not at all unusual under the circumstances. However, veteran astronaut John Young, who was then participating in his fifth spaceflight, had a heartbeat that hovered around 85 beats per minute. "I was excited too," Young admitted, "I just can't make it go any faster!" This onboard photo of Young during the mission was taken by Crippen. His three-piece constant-wear flight suit is a far cry from the bulky suits of the early days.

> **Yoda Unavailable Was.** NASA once considered selecting a midget as the first American in space in order to reduce the payload weight of the rocket and capsule.

Q *Name the astronaut who was at first rejected by NASA as being too tall at an even 6 feet.*

A Thomas Stafford, who was 1 inch above the maximum. He became an astronaut in NASA Group 2, which included nine new spacemen.

Q *Which astronaut has flown the most missions?*

A John Young, six through December 1984.

Q *Which of the Original 7 Astronauts had compiled the most time flying jet planes?*

A Gus Grissom, who flew more than 100 combat missions in F-86 Sabrejets during the Korean War.

Q *How old was Soviet cosmonaut Yuri Gagarin when he made the historic flight in Vostok I?*

A Twenty-seven years old.

Q *Name the oldest space traveler.*

A Dr. William Thornton, who was 54 when he flew STS-8 in August 1983. He took the record from Deke Slayton, who was 51 when he flew the Apollo-Soyuz mission in 1975.

Q *What special lunar activity support equipment was included aboard Apollo 14?*

A A cane, for the "elderly" Alan Shepard, then 47. The gag, with an appropriate label, came from pad manager Guenter Wendt.

Q *Why did astronauts often have fluctuating heart rates before launch?*

A Because pad manager Guenter Wendt would mount photos of nude women in front of the observation port the crew members looked through during launch preparations. Heart and blood pressure readings often changed abruptly upon discovery.

Q *What did Wally Schirra find when he boarded his Mercury spacecraft?*

A Car keys, dangling from the controls.

Q *Which astronaut waited the longest for a flight?*
A Bruce McCandless. Selected in 1966, he didn't fly until February 1984, almost eighteen years later. Another Group 5 astronaut, Don Lind, is slated to fly in mid-1985. He will have waited nineteen years. McCandless is seen here with his crewmates of STS-11, clockwise from the top: Ronald McNair (standing), McCandless, Robert Gibson, Vance Brand, Robert Stewart.

> **Job Specifications** NASA's original job specifications for astronaut candidates did not state that the applicants had to be pilots. Though pilots were among those who were eligible, the list also encouraged applications from parachutists, submariners, mountain climbers, Arctic explorers, deep-sea and scuba divers, combat veterans and people with combat training, and persons who had been test subjects in acceleration and atmospheric-pressure tests. The reasoning was that the person riding in a space *capsule* would not have to control it the way a pilot does.

Q *Name the first astronaut to resign.*

A John Glenn. Seeing little chance of being given another flight, he became the first to parlay astronaut glitter into corporate gold. Glenn took an executive spot with Royal Crown Cola for a reported $50,000 salary plus stock options—pretty good money back in 1964. He also retired from the U.S. Marine Corps at the same time, in order to run for a U.S. Senate seat from Ohio. Hardly a month later he was injured in a fall and was forced to drop out of the race for the Democratic nomination.

Q *Which astronaut spent the shortest total time in space after reaching orbit?*

A John Glenn: 4 hours, 55 minutes. (The longest was spent by the Skylab 4 crew—Gerald Carr, Ed Gibson, and William Pogue—in 1973–74.)

Q *How long did Alan Shepard's Project Mercury flight last?*

A The suborbital flight, which reached an altitude of 115 miles, lasted 15 minutes.

Q *Identify the first manned mission to last longer than 200 days.*

A A 1982 Salyut mission, flown by Anatoly Berezovoy and Valentin Lebedev, which lasted 211 days. Lebedev would have been on a 185-day mission in 1980, but he broke his leg on a trampoline just days before launch and was replaced by Valery Ryumin, who had just recently returned from a 175-day mission.

Q *Identify the two spacecraft involved in the first rendezvous in flight.*

A Gemini 6A, with Wally Schirra and Thomas Stafford (December 15–16, 1965), and Gemini 7, with Frank Borman and Jim Lovell (December 4–18). Schirra maneuvered Gemini 6A to within 6 inches of Gemini 7. The two spacecraft flew in the same orbit for 5½ hours, and at no time was the distance between them greater than 100 feet.

Seeing Things?—Among the initially unidentified objects seen in space by astronauts were the "beer can" reported by James McDivitt on Gemini 4, later identified as part of a booster. John Young and Michael Collins reported "red lights" shortly after reaching orbit on Gemini 10. While Collins prepared for his spacewalk, Young reported seeing an extremely bright object through his porthole; it was later identified as Venus. McDivitt is seen here having his eyes examined prior to flight test selection.

> **Take Me to Your Leader** During his debriefing after the flight of *Friendship 7*, John Glenn spoke at some length about the tiny luminescent particles he saw during all three orbits. The particles were quite a mystery and seemed to intrigue Glenn. NASA staff psychiatrist Dr. George Ruff heard Glenn describe the particles and posed the following question: "What did they say, John?"

Q *What is the Glenn Effect?*

A A swarm of small luminous particles surrounding an orbital craft. First reported by John Glenn, they were later reported by Scott Carpenter, who identified them as ice crystals and flecks of oxidized paint from the external hull.

Q *How many astronauts claim to have seen UFOs in space?*

A UFO sightings are attributed to at least five nauts: Scott Carpenter, Gordon Cooper, John Young, Gene Cernan, and James McDivitt. However, none of them thought the sighting significant enough to file the required UFO report.

Q *How did the Soviets use UFOs to hide the existence of a second launch site?*

A When thousands of Russians saw the flames and contrails from a launch from Plesetsk in the northwestern U.S.S.R., the government called it "glowing industrial pollution." This is the Soviet version of the "swamp gas" excuse, and it convinced many that the Kremlin was covering up a UFO sighting.

Q *Which astronaut crew gave the most detailed UFO sighting?*

A Wally Schirra and Tom Stafford of Gemini 6, on the morning of December 16, 1965. They described a large object flying below them on a north-to-south trajectory, and even said they were worried about possible collision. After describing the lights and shape of the UFO in great detail to a worried Mission Control, they broke out a set of bells and a four-hole harmonica and treated the world to "Jingle Bells." Gotcha!

Q *Has the government ever investigated what effect the discovery of extraterrestrial life might have on society?*

A In the 1960s NASA and the Brookings Institute did such a study and concluded that there would be no significant effect if the life-form is of subhuman intelligence. But if "they" are smarter than us, it could upset many philosophies and value systems—and would be most devastating to scientists and engineers!

Generally Speaking—Michael Collins, who flew in space in Gemini 10 and was part of the Apollo 11 crew, is the son and nephew of two U.S. Army generals. His father was Maj. Gen. James J. Collins and his uncle was Gen. J. Lawton ("Lightening Joe") Collins. Mike Collins himself is a 1952 graduate of West Point. Here a pensive Collins studies a lunar map in preparation for the historic Apollo 11 flight.

Late Bloomer   A high-school dropout, Story Musgrave enlisted in the Marines. By the time he came to NASA with Group 6, he had finished high school and had acquired a B.S. in statistics, a B.A. in chemistry, an M.B.A. in operations analysis and computer programming, an M.S. in biophysics, and an M.D.—and he was working on a doctorate in physiology and continuing an internship in surgery. He became the second American M.D. in space, on STS-6.

Q *Which of the Original 7 Astronauts dropped out of college in his junior year but later was awarded an honorary degree from the same school?*

A John Glenn, who left Muskingum College in Ohio.

Q *What was the only elective course Jim Lovell took at Annapolis?*

A Allowed only one elective, a choice of language, Lovell chose German so he could read the works of von Braun and Oberth.

Q *What was the first experiment sent into space by a high-school student?*

A Eighteen-year-old Todd Nelson of Rose Creek, Minnesota, devised an experiment to check the effects of weightlessness on insect flight. Dozens of moths, houseflies, and honeybees were orbited on STS-3: they walked.

Q *Have living organisms ever been discovered beyond the Earth?*

A Apollo 12 astronauts Conrad and Bean found part of Surveyor 3 on the Moon in November 1969. In the TV camera's foam insulation was discovered a colony of bacteria, *Streptococcus mitis*. The germs had survived lunar conditions for 2½ years.

Q *Which astronaut claimed he got lost every time he went to the Pentagon?*

A Mike Collins, who performed successful celestial navigation on Gemini 10 and Apollo 11, within 0.01% accuracy in one case.

Q *How were the giant parts of the Saturn rocket transported from their assembly plants to Cape Kennedy?*

A A special trailer and barge were required for moving the 138-foot-long first stage over land and water. The second stage was also transported by ground and water vehicles. Only the third stage could be transported by plane, and this required the gigantic Super Guppy.

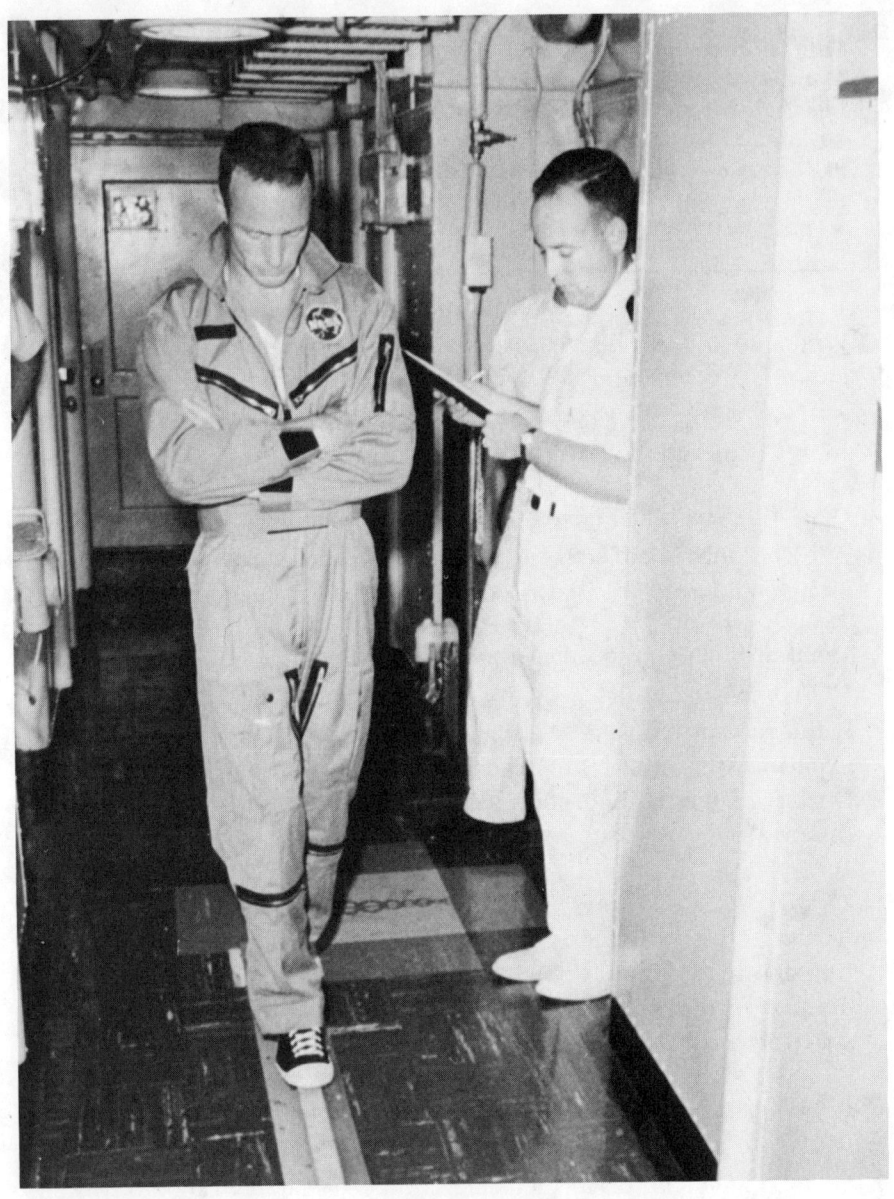

But Watch That Third Rail!—After his spaceflight, John Glenn had to walk along a series of rails as a test of his balance. Some time later Col. Chuck Yeager, the man who broke the sound barrier, told Glenn that the test only proved that Glenn had been brought up near railroad tracks! In this photo, Scott Carpenter is seen "walking the rail" after his flight.

Purdue Claims First, Last Moon Footprints Both the first man to set foot on the Moon, Neil Armstrong, and the last, Gene Cernan, are graduates of Purdue University. Gus Grissom and Roger Chaffee were also Purdue graduates, and no fewer than seven other Purdue alumni were members of the Astronaut Corps as of March 1984: John Blaha, Roy Bridges, Richard Covey, Guy Gardner, Jerry Ross, Loren Shriver, and Donald Williams.

Q *Which astronaut did ESP experiments en route to the Moon?*

A Edgar Mitchell, during Apollo 14. He carried a deck of twenty-five symbol cards, and during rest periods he attempted to send impressions of the cards to four psychics back in the U.S. The best score on the receiving end was 51 out of 200. Random guessing turns up 40 out of 200, so the findings were considered to be of little significance.

Q *Who was the first American to ride an Atlas rocket in a spacecraft?*

A John Glenn, during the flight of Friendship 7. Alan Shepard and Gus Grissom were launched by Redstone rockets for their suborbital flights.

Q *How many suborbital flights were originally scheduled?*

A Seven, one for each of the astronauts. Only Shepard and Grissom, however, went suborbital. The remaining five flights were canceled as unnecessary.

Q *How fast did the Mercury spacecraft travel?*

A The two suborbital flights reached top speeds of 5,200 mph. The orbital flights made it to 17,500 mph.

Q *Where does "sky" end and "space" begin?*

A There's still measureable atmosphere at 140 miles, but early on American space officials set an arbitrary "border" at 50 miles.

Q *Name the first human to fly in space twice.*

A Gus Grissom, who soloed in the second Project Mercury flight and was part of the two-man crew in the first Gemini flight. His *Liberty Bell 7* was launched on July 21, 1961, and he and John W. Young lifted off in Gemini 3 on March 23, 1965.

Q *Name the members of the second group of astronauts.*

A Group 2, selected on September 17, 1962, included (clockwise from top right): Frank Borman, John Young, Thomas Stafford, Pete Conrad, James McDivitt, James Lovell, Elliot See, Edward White, and Neil Armstrong.

The Long and Short of It The night before the preselection physicals for Group 3, C. C. Williams stayed awake and bounced up and down on his heels. It may sound like a frat hazing, but the 6-foot 1-inch pilot realized that the bouncing might compress his spine enough to squeak by the 6-foot height limit. It worked.

Q *Name any seven of the fourteen astronauts in Group 3.*

A Selected in October 1963, they were: Buzz Aldrin, Bill Anders, Charles Bassett, Alan Bean, Gene Cernan, Roger Chaffee, Mike Collins, Walt Cunningham, Donn Eisele, Ted Freeman, Dick Gordon, Rusty Schweickart, Dave Scott, C. C. Williams.

Q *Of the first sixteen astronauts (the Original 7 and the Next 9), name the only one to hold two world aircraft speed records.*

A John Young, selected in the second group. John Glenn held one world speed record.

Q *Name the first astronaut to spend a full day in space.*

A Gordon Cooper, aboard *Faith 7*, May 15, 1963. His flight lasted 34 hours, 19 minutes, and 49 seconds and completed 22.5 orbits of the Earth, the most orbits of any of the Project Mercury flights.

Q *Name the first person to make actual physical contact with another orbiting object in space.*

A Michael Collins, during Gemini 10. During his spacewalk Collins touched the Agena 8 satellite, which had been put into space some three months earlier. The flight also marked the first dual rendezvous in space, since Gemini 10 had earlier docked with its own Agena for 39 hours.

Q *Who remained in* Columbia *when the* Eagle *landed on the Moon?*

A Michael Collins was command module pilot and remained on board while lunar module pilot Buzz Aldrin and commander Neil Armstrong took the *Eagle* to the surface of the Moon in July 1969.

Q *What was unusual about the voice of the cosmonaut aboard Sputnik 10?*

A "Ivan Ivanovich," the Russian version of John Doe, was a dummy and was rigged to play tapes of the Pyatnitsky chorus for the benefit of Western eavesdroppers.

Q *How many Americans have walked on the Moon, and how many can you name?*

A Twelve. In order: Neil Armstrong and Buzz Aldrin (Apollo 11), Pete Conrad and Alan Bean (Apollo 12), Alan Shepard and Edgar Mitchell (Apollo 14), Dave Scott and Jim Irwin (Apollo 15), John Young and Charles Duke (Apollo 16), Gene Cernan and Jack Schmitt (Apollo 17).

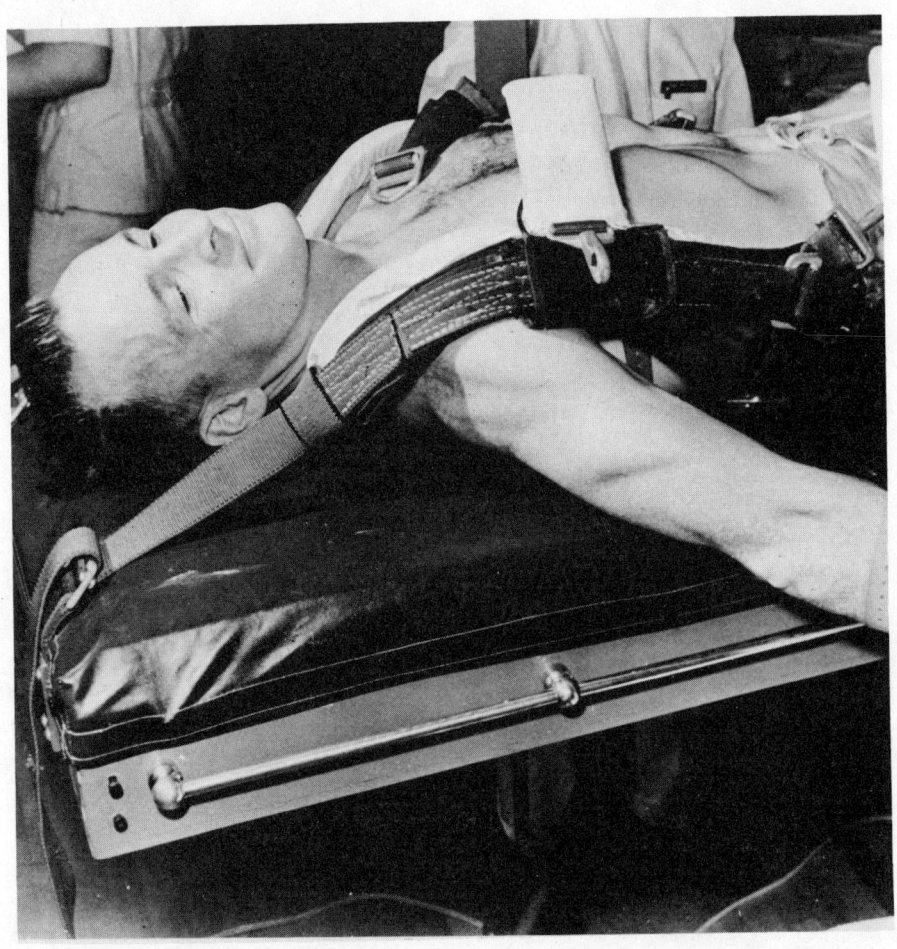

Q *Name the first American to walk in space.*

A Ed White, aboard Gemini 4, on June 3, 1965. He spent 23 minutes outside the craft, floating at the end of a 24-foot-long gold-plated umbilical. White is seen here as his blood-pressure reflex control is measured prior to his selection as a member of flight test personnel.

See Ya Around, Chum   What would have happened if a spacewalking Gemini astronaut became "incapacitated"? In the event of unconsciousness or death, if the onboard astronaut could not drag the victim back into the spacecraft and securely seal the hatch (unlikely with the bulky pressure suits), he was to cut the hapless partner loose and land by himself.

Q *Who was the first man to walk in space?*

A Alexei Leonov, during the March 18, 1965, flight of Voskhod 2. Crewmate Pavel Belyayev was positioned to manually "reel in" Leonov if necessary. The spacewalk lasted a total of 20 minutes.

Q *Why did some Americans doubt the authenticity of the Russian spacewalk?*

A The film of the extra-vehicular activity (EVA) released by the Soviets included some studio footage, added for dramatic effect and better visibility. These inserts were not clearly identified as such and led to Western debunking of the feat as a hoax. Published data later supported the Soviet claims.

Q *How did America's first spacewalker describe his feelings afterward?*

A Gemini 4's Ed White, who made the walk on June 4, 1965, said, "I felt red, white, and blue all over."

Q *What problems almost killed the first spacewalker?*

A When Alexei Leonov tried to re-enter the Voskhod 2 spacecraft on March 18, 1965, he was unable to maneuver his bulky spacesuit into the narrow hatch. After several tiring attempts, it was decided that if the suit was partially deflated, there might be enough flexibility in the joints to make it. Although the depressurization exposed Leonov to the bends, it was better than being cut loose. Gemini spacewalkers reported similar problems.

Q *Identify the only one of the Original 7 Mercury Astronauts to walk on the Moon.*

A Alan Shepard, during Apollo 14, launched on January 31, 1971.

Q *Who was the first astronaut to step outside an orbiting spacecraft twice during the same flight?*

A Michael Collins, who did it during the flight of Gemini 10.

Q *Who made the first spacewalk from a shuttle?*
A Story Musgrave, followed shortly by Donald Peterson, both seen here on the STS-6 mission, 1983.

Q *Name the first man to make an untethered spacewalk.*

A Bruce McCandless, on STS-11, February 4, 1984, shown here flying with the manned maneuvering unit, a backpack rocket system. McCandless flew over 100 meters from the *Challenger* at a top speed of 7.2 feet per second.

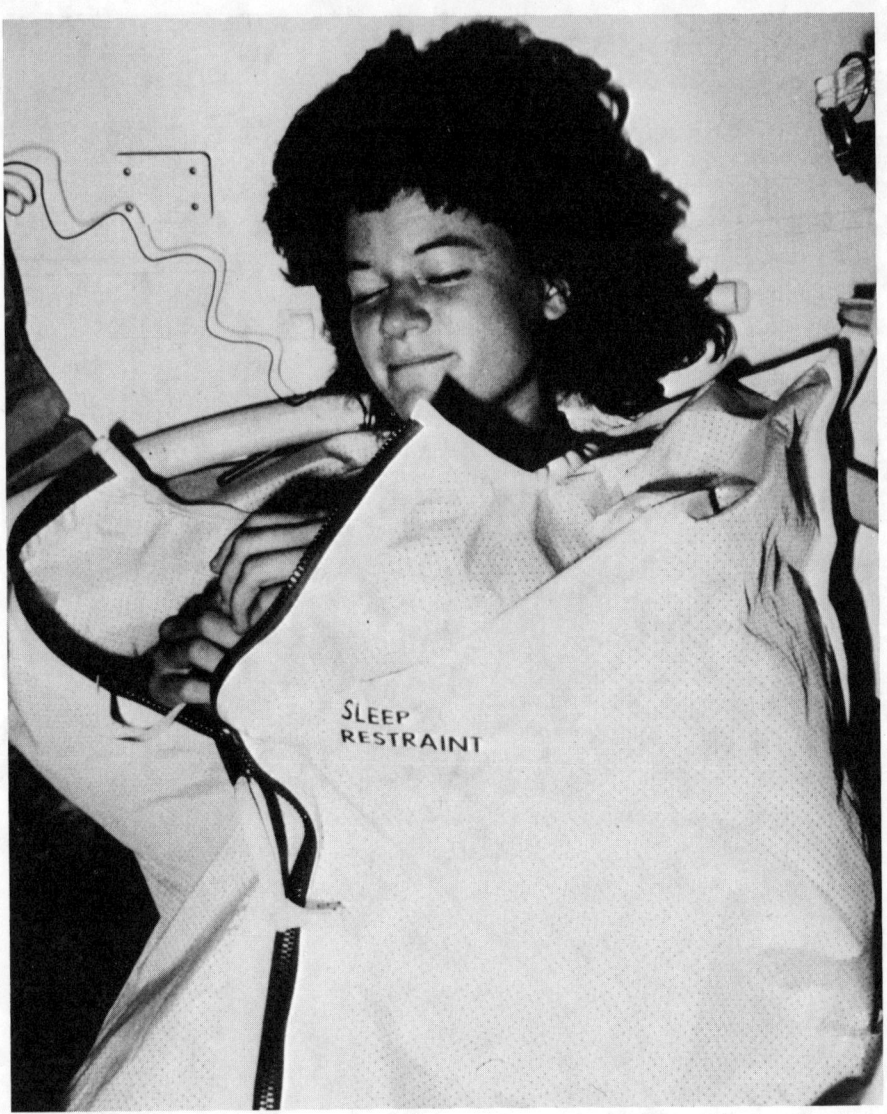

SLEEP
RESTRAINT

Space Sleepers—During the early planning stages of Project Mercury serious consideration was given to a plan to anesthetize the astronauts who would ride into space. The idea was that in this immobile state they could not tamper with the controls of the spacecraft or do anything else that might disturb the scientific evaluations being made. In addition, it was thought that readings from body sensors on the astronauts would be less prone to changes if the astronauts panicked. By the time Sally Ride rode into space aboard STS-7, she and other astronauts had the option of using a sleep restraint for real sleep purposes.

> **Just One of the Boys?** When the first group of female astronaut candidates arrived at Johnson Space Center in 1978, it took NASA a little while to realize that some changes would be in order. No arrangements had been made for in-flight sleepwear (males had slept in their underwear or less), deodorant selection was limited to Old Spice and English Leather, and tampons had not been considered.

Q *Identify the astronaut who actually fell asleep in his spacecraft during one of the "holds" prior to launch.*

A Gordon Cooper, in *Faith 7*, the last Project Mercury flight, which was launched on May 15, 1963.

Q *When were U.S. women first tested as astronaut candidates?*

A In 1959. A group of thirteen women began medical testing in New Mexico. The project was kept secret for three years, until after it was dropped. They called themselves FLATs, for First Lady Astronaut Trainees.

Q *How did the female astronauts' test results compare with those of the all-male Project Mercury astronauts?*

A They were generally better. Women sustained longer periods of sensory deprivation; withstood pain, g-force, heat, cold, and noise better than the men; and were more resistant to heart problems and radiation. What's more, they were generally smaller, consuming less air, water, and food, taking up less space, and adding less weight.

Q *Who was the first woman in history known to have applied for spaceflight?*

A In 1927, German scientist Max Valier was quoted in newspapers as saying that a lunar flight would be possible within a few years. A Russian woman named Olga Vinnitskaya wrote to Tsiolkovsky, asking him to intercede with Valier so that she might be considered for the mission. Tsiolkovsky told her the story had been exaggerated.

Q *What was NASA's reason for not sanctioning the women's program?*

A The official reason was that the FLATs weren't *military* test pilots and had no experience with supersonic jets; this was a Catch-22 since the military wasn't accepting female pilots at the time. Other astronauts have suggested the real reason was that public sentiment would have been offended by overnight orbital mixed cohabitation, a problem that could have been solved by the even more unthinkable all-female crew.

Q *How many married astronaut couples are there?*

A Three. Drs. Bill and Anna Fisher were married before they entered the program. Robert Gibson and Margaret Rhea Seddon met in the program and married after completing training, as did Steven Hawley and Sally Ride. The Fishers were photographed together during a weightless moment in a NASA DC-135 aircraft while training in November 1980.

> **Really a Wrong Number!** Kennedy Space Center had a sexy-voiced phone operator known only by her number. Many astronauts were captivated by her throaty, seductive conversation, but the few who followed up found that, in the words of one space jock, "she was as sexy as Martha Raye in Army fatigues." Old hands would sucker newcomers into making a date with her, sight unseen.

Q *Can you name any of the FLATs?*

A Out of the twenty-five applicants, thirteen passed the first medical and psychological screening: Rhea Hurrle Allison, Myrtle Cagle, Jerrie Cobb, Jan Dietrich, Marion Dietrich, Mary Wallis Funk, Sarah Lee Gorelick, Jane Hart, Jean Hixson, Irene Leverton, Geraldine Sloan, Bernice Trimble Steadman, and Gene Nora Stumbough.

Q *Did NASA ever consider using real twins in Project Gemini?*

A Yes. Two women, Jan and Marion Dietrich, were said to be under consideration in the early 1960s. Some thought twins could work together better than nontwins, and could better cope with the stresses of long-duration flight.

Q *Who was the only FLAT to work for NASA after the project was disbanded?*

A Jerrie Cobb, the most prestigious test pilot of the group. She worked as a consultant.

Q *What is the name of the first child born to parents who had both flown in space?*

A Yelena Nikolayeva. Her father was Soviet Air Force Maj. Gen. Andrian G. Nikolayev of Vostok 3 and Soyuz 9, and her mother was Soviet Air Force Col. Valentina V. Tereshkova of Vostok 6. The couple were married in November 1963, and Yelena was born seven months later in June 1964. When Khrushchev learned of their plans to marry, he asked them to wait, lest it appear they had been ordered to marry and have children to test the effect of spaceflight radiation on reproduction. When it was explained why they couldn't wait, Khrushchev made the best of it and threw the biggest wedding Russia had seen since Nicholas and Alexandra's.

Q *Who was named National Father of the Year in 1962?*

A John Glenn. His mother was selected World Mother of the Year.

Q *How many astronauts have been elected to Congress?*

A Three. John Glenn as senator from Ohio (1974–), Harrison Schmitt as senator from New Mexico (1976–82), and Jack Swigert as congressman from Colorado (1982). Swigert succumbed to cancer in December 1982, before he could be sworn in. Seen with Glenn at the launch control center, prior to launch of Apollo 4, are two other astronauts whose names have been linked to politics: Jim Lovell (left) was frequently mentioned as a candidate from Wisconsin, and Al Worden (right) lost a race for Congress in Florida in 1982. Meanwhile, both Texas and Virginia sought to have Alan Shepard run for U.S. Senate seats, Charles Duke was approached in North Carolina, Don Lind in Utah in 1971, and Tom Stafford considered a U.S. Senate seat from Oklahoma. Jack Lousma lost a bid for a U.S. Senate seat from Michigan in 1984.

Flag Day   Astronaut Ed White presented U Thant, secretary general of the United Nations with the small blue and white U.N. flag he had carried with him during his Gemini 4 spacewalk. It was the same flag White's father had carried while a member of U.N. forces during the Korean conflict.

Q *Identify the first astronaut to address the United Nations.*

A John Glenn, on March 2, 1962. The day before, he and his fellow astronauts had been given a ticker-tape parade in New York City before a crowd esimated at more than 4 million. During a brief ceremony at City Hall Plaza after the parade, the city's Department of Sanitation Band performed.

Q *Which astronaut became a delegate to the U.N.?*

A Alan Shepard served as representative to the General Assembly in 1971, appointed by President Nixon.

Q *Which astronaut became a U.S. ambassador?*

A William Anders, who flew around the Moon on Apollo 8. He was named Ambassador to Norway in 1972. (His ancestors came from Norway.)

Q *Which astronaut went on to manage the New Orleans Saints?*

A Dick Gordon.

Q *Which astronauts went on to become Coors beer distributors?*

A Al Shepard and Stu Roosa.

Q *With what spacecraft project did Deke Slayton become associated after leaving NASA?*

A The Conestoga rocket, a privately produced expendable vehicle.

Q *Which astronaut took part in a march following the murder of Dr. Martin Luther King?*

A Buzz Aldrin. While he was never called on the carpet for it, the move was highly controversial in conservative Texas.

Q *What was the very last piece of Gemini hardware to be certified as functional prior to the first manned mission?*

A The astronaut ejection escape system.

Q *Which astronaut quit to devote full time to his artwork?*

A Painter Alan Bean, the fourth Moonwalker. However, in October 1969, the Apollo 12 lunar module pilot's thoughts were still on the scientific experiments he and Pete Conrad would be conducting.

Q *Which astronaut has made several expeditions to search for Noah's Ark?*

A James Irwin, who heads the High Flight Foundation, a Colorado Springs ministry. He has made trips to Mt. Ararat in Turkey which have been adventurous but inconclusive. Another mountain, Hadley Rille on the Moon, was the site of more productive excursions for Irwin, Al Worden, and Dave Scott. Here Irwin is seen bending over the lunar rover during its debut on Apollo 15 in August 1971.

Q *Besides the Cape Canaveral area, what other sites were considered for Moonports?*

A Cumberland Island, Georgia; Mayaguana, in the Bahamas; White Sands Missile Range, New Mexico; South Point, Hawaii; Christmas Island, in the South Pacific. The July 1961 study also considered building an artificial island east of Florida as a base for the giant Saturn 5. Despite fear of Defense Department domination, NASA decided to build what is now the Kennedy Space Center near existing Air Force facilities, Cape Canaveral and Patrick Air Force Base.

Q *What problems with the design of the Gemini spacecraft were unconsciously created by Gus Grissom?*

A Grissom, forgetting that other astronauts might be larger than his own 5-foot 7-inch frame (he was the shortest of the Original Seven, however, he had the longest torso), had the interior of the spacecraft designed in a way that was comfortable for himself. Tom Stafford was part of the backup crew for Grissom and Young's flight in the first manned Gemini, and at an even 6 feet, he was uncomfortable in practice simulations. Modifications were eventually made. In the photo above, a Mercury and a Gemini are shown side by side.

Q *Match the ship or ships on the left with the missions on the right.*

a. USS *Princeton*
b. USS *Guadalcanal*
c. USS *Wasp*
d. USS *Intrepid*
e. USS *Pierce*
f. *Twin Falls Victory*
g. *Ranger Tracker*
h. *Rose Knot*
i. *Coastal Sentry*
j. *Huntsville*
k. *American Mariner*
l. *Watertown*

1. Communications ship stationed in the Indian Ocean
2. Was added to ground communications network for Mercury's *Sigma 7* and stationed off Midway
3. Recovered the crew of Gemini 10
4. Recovered Scott Carpenter
5. Was added to Mercury's *Faith 7* ground communications network and stationed in the Atlantic
6. Picked up the last manned flight before a lunar landing
7. Made the most recoveries of crews during Gemini program
8. Was added to Mercury's *Faith 7* ground communications network and stationed off the Gilbert Islands
9. Recovered Scott Carpenter's spacecraft, *Aurora 7*
10. Communications ship stationed in the Atlantic Ocean

A a-6 USS *Princeton* picked up the crew of Apollo 10 on May 26, 1969.
b-3 USS *Guadalcanal* recovered John Young and Michael Collins on July 21, 1966.
c-8 USS *Wasp* recovered the crews of Gemini 4, 6, 7, 9, and 12.
d-4 USS *Intrepid* recovered Scott Carpenter in *Aurora 7*.
e-9 USS *Pierce* recovered Scott Carpenter's spacecraft, *Aurora 7*, two hours after USS *Intrepid* picked up Carpenter.
f-5 *Twin Falls Victory* was stationed between Bermuda and Florida during Gordon Cooper's *Faith 7* Mercury flight.
g-8 *Ranger Tracker* had the same communications duty as *Twin Falls Victory* off the Gilbert Islands in the Pacific during *Faith 7*.
h-10 *Rose Knot* was a communications ship in the Atlantic.
i-1 *Coastal Sentry* had the same duty as *Rose Knot*, but her station was the Indian Ocean.
j-2 *Huntsville* was one of three ships added to the ground communications network for Wally Schirra's *Sigma 7*.
k-2 See j-2. *American Mariner* was the second ship added to this mission.
l-2 See j-2. *Watertown* was the third ship added to this mission.

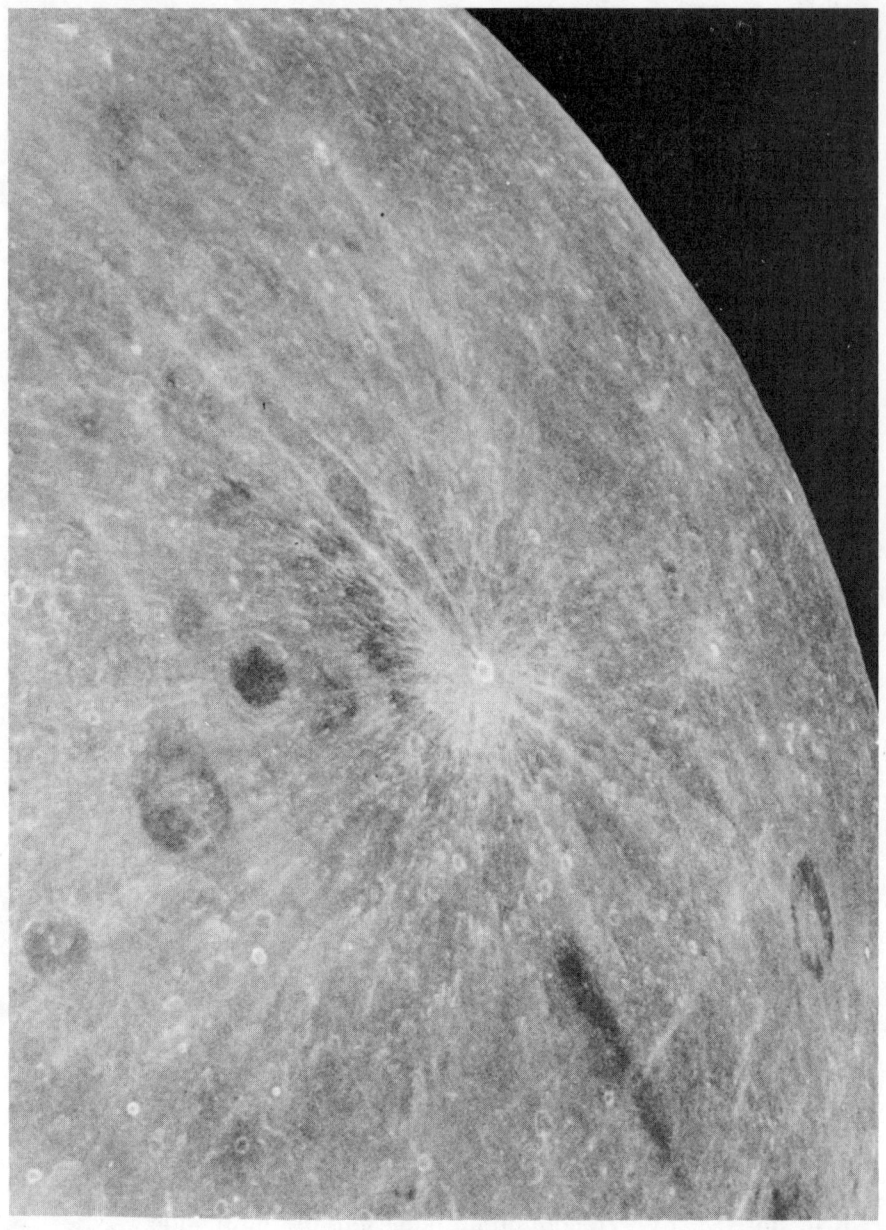

"I Don't Want to Set the Moon on Fire"   There was concern among geologists that the lunar soil or rocks might contain pure metallic elements, which would burst into flames when brought into the pure-oxygen environment of the LM, even just in the form of dust the astronauts may have tracked back in on their boots. They didn't.

> **A Tall, Cool One**   Because the fuel propellants in a Saturn rocket had to be kept at temperatures below 200° F, there was a buildup of 1,400 pounds of ice on the outer skin of these rockets prior to launch.

Q *Identify the U.S. destroyer that recovered John Glenn and his spacecraft after his historic flight.*

A USS *Noa,* whose number designation was DD-841 and whose nickname was "Steelhead." Glenn was transferred by helicopter to the aircraft carrier USS *Randolph.*

Q *Identify the U.S. aircraft carrier that recovered the first unmanned Mercury-Redstone (MR) spacecraft to land in the sea.*

A The USS *Valley Forge,* on December 19, 1960.

Q *Identify the U.S. Navy ship that recovered the chimpanzee Enos after his orbital flight in MA-5.*

A The USS *Stormes,* which lifted the spacecraft out of the water slightly more than an hour after splashdown in the Atlantic.

Q *Identify the aircraft carrier that recovered the crew of the first manned Apollo mission.*

A The USS *Essex.* It recovered Wally Schirra, Donn Eisele, and Walt Cunningham after Apollo 7 splashed down within 8 miles of the ship on October 22, 1968.

Q *Name the only U.S. ship to recover two of the Project Mercury astronauts.*

A The USS *Kearsarge,* which recovered Wally Schirra and Gordon Cooper. John Glenn was recovered by the USS *Noa* and then transferred to the USS *Randolph.* The *Randolph* recovered Grissom. Alan Shepard was recovered by the USS *Lake Champlain* and Scott Carpenter by the USS *Intrepid.*

Q *Identify the launchpad, by number, that Alan Shepard lifted off from.*

A Pad 5, Cape Canaveral, which NASA technicians called Surfside 5, after the popular TV show *Surfside 6.*

Q *What was the number of the pad that John Glenn's flight was launched from?*

A Pad 14.

Do You Have an Account Here?—NASA provided personal kits for the astronauts on the recovery ships, so they would have some necessary items after being fished out of the sea. The kit John Glenn received aboard the USS *Noa* included underwear, dark glasses, a watch, a flying suit, sneakers, and a blank check. "I assumed that the check was to be made out to my own bank account to provide me with some spending money until I could get home," he said. Glenn is seen here aboard the *Noa*, 21 minutes after splashdown off Grand Turk Island.

> **The Price of Fame** At one point, Gus Grissom computed the hours he had actually spent on the space program. Based on his USAF captain's pay, he figured that being an astronaut paid just slightly less than $1 per hour.

Q *How much was Alan Shepard paid for being the first American in space?*

A Flight pay for the 15-minute trip came to $14.38.

Q *How much did John Glenn make?*

A His pay was $245 . . . less than a penny a mile.

Q *How much did it cost us to go to the Moon?*

A The total cost for the manned space program, from Project Mercury through Project Apollo, was about $45 billion. By comparison, the Soviets spent $85 billion during the same period.

Q *Why was Neil Armstrong's expense allowance cut during his mission to the Moon?*

A A GS-16 civil servant normally gets $16 per day for expenses while traveling out of town. But since Armstrong had government-supplied food and lodging, it was cut to $2 per day. Armstrong says he netted $49.10 for the week-long mission and pre-flight period at KSC.

Q *Which astronaut asked for an extension on his income tax deadline because he was on his way to the Moon?*

A Jack Swigert, who was assigned to go to the Moon only a week before the April 15 deadline. An extension is normally provided for "citizens who are out of the country." When the returning crew arrived at American Samoa in the Pacific, Gov. John Haydon had a 1040 form ready for Swigert.

Q *How many people were on the Project Mercury payroll when John Glenn made his orbital flight aboard* Friendship 7 *on February 20, 1962?*

   a. *under 1,000*   b. *5,000*   c. *over 10,000*

A Just under 1,000. Fifteen months later, when Gordon Cooper rode *Faith 7* into orbit, Project Mercury had dwindled to fewer than 500 people. By that time the bulk of NASA personnel were working on projects Gemini and Apollo (over 7,000 on Apollo alone).

Q *Did any astronauts win Emmy awards?*
A Wally Schirra, Donn Eisele, and Walt Cunningham won a special award for making the first live network TV broadcast from space during the flight of Apollo 7, October 11–22, 1968.

Money Well Spent In July 1967, as the race to the Moon swung into high gear, NASA paid a record $7.3 million in overtime alone.

Q *Name the first live-on-TV coverage of a spacecraft returning to Earth (including the parachute descent).*

A The March 13, 1969, recovery of Apollo 9, in the waters off Grand Turk Island in the Bahamas. They splashed down within 3 miles of the recovery ship, the USS *Guadalcanal.*

Q *Name the longest-duration spaceflight through the end of 1984.*

A The Salyut 7/Soyuz T-10 mission of Leonid Kizim, Vladimir Solovyev, and Oleg Atkov, 1984; 237 days.

Q *Which Apollo mission transmitted the first color photographs of the full Earth?*

A Apollo 10, May 18–26, 1969. People on Earth had an opportunity to see our planet exactly as it appeared to the astronauts.

Q *When did the first live TV broadcast of a Soviet launch take place?*

A On July 15, 1975: the launch of Soyuz 19, which linked up with an Apollo spacecraft. The only other live launch to date was the April 1984 launch of Soyuz T-11, which carried India's first space traveler, Rakesh Sharma. The joint U.S.-Soviet mission was also the first time the Russian crew was identified before launch, and the first time foreign observers were allowed at Tyuratam.

Q *Which Soviet mission included the first live TV transmission?*

A Vostok 2. Asked if American officals had any doubt that the Russians had really accomplished the feat, Wernher von Braun told reporters that NASA had found the Soviet telemetry frequency and had "watched his face" during the flight. That information was supposed to be secret.

Q *Which astronauts saw Batman on their onboard TV monitor?*

A Pete Conrad and Dick Gordon, on Gemini 11. The picture of Batman and Robin was actually placed over the monitor as a prelaunch gag.

Q *Which astronaut spent two weeks on a TV game show?*

A John Glenn, as a contestant on *Name That Tune* in 1957.

Q *Name the first black American in space.*

A Lt. Col. Guion Bluford, above, who served as a mission specialist on STS-8, August 30, 1983. While he was the first black American in space, he was neither the first black astronaut nor the first black in space, as some reports claimed.

> **More Than Just Astronauts** At the height of the space program, Project Apollo employed just over 410,000 people.

Q *Name the man the White House touted as the "First Negro Astronaut."*

A Air Force captain Ed Dwight. Because the Kennedy Administration let it be known in both NASA and the Air Force that it wanted at least one Negro selected as an astronaut, Dwight was groomed for the job, but like many other qualified pilots, he failed to make the Air Force cut when the top eleven were chosen from the twenty-five pilots in training. In order to satisfy the White House request, the "paper" cut was expanded to the top fourteen, which included Dwight. The White House leaked stories to the press hinting that Dwight would be the first Negro astronaut, but he was eliminated in Air Force screening before ever being considered by NASA.

Q *Who was the first black astronaut trainee?*

A Maj. Robert Lawrence, Jr., selected June 30, 1967, as part of the Air Force Manned Orbiting Laboratory project. A chemist as well as a test pilot, Dr. Lawrence died in the crash of an F-104 at Edwards in December of the same year. NASA's first black astronaut candidates were Lt. Col. Guion Bluford and Ronald McNair, both selected in Group 8, January 1978.

Q *Who was the first black in space?*

A Arnaldo Tamayo-Mendez of Cuba, who was also the first Hispanic. The second black in space was Guion Bluford of the U.S.

Q *How many American astronauts were born outside the U.S.?*

A Five: Mike Collins (Rome, Italy); Bill Anders (Hong Kong); Franklin Chang (San Jose, Costa Rica); Philip Chapman (Melbourne, Australia); Shannon Lucid (Shanghai, China).

Q *Who was the first non-WASP American in space?*

A Wally Schirra, of Swiss ancestry.

Q *Which astronaut later went into the wasp-breeding business?*

A Scott Carpenter.

Q *Who was the first civilian astronaut to fly?*

A Neil Armstrong, on GT-8 with Dave Scott.

Q *Who was the first non-American to fly on an American space mission?*
A Discounting the Apollo-Soyuz Test Project of 1975, the first was West German scientist Dr. Ulf Merbold on STS-9, November 1983, shown here aboard Spacelab.

Rising Son (or Daughter) of Rising Sun? The Japanese space agency has received applications from more than 550 men and women who hope to be among the 3 to be chosen for NASA training for a shuttle flight. They will train for two years, and one will be a crew member aboard STS-68, now scheduled for 1988.

Q *Who was the first Oriental in space?*

A Pham Tuan of Vietnam. He was followed by Jugderdemidin Gurrugcha of Mongolia and Franklin Chang of the U.S.

Q *Besides the United States and the Soviet Union, what nations are training astronauts?*

A Payload specialists from West Germany, Australia, Canada, Switzerland, and the Netherlands have been trained by NASA for Spacelab flights, and the Chinese have reportedly begun training for manned flights.

Q *Who was the first non-American, non-Soviet man in space?*

A Vladimir Remek of Czechoslovakia. He flew with two Russians on Soyuz 28, on March 2, 1978. While "guest cosmonauts" from communist nations did a few experiments, they did little serious research.

Q *Who was the first non-American, non–Soviet-bloc man in space?*

A Jean-Loup Chrétien of France, who flew on Soyuz T-6 with Vladimir Dzhanibekov and Alexander Ivanchenkov in June and July 1982. They docked with the Salyut 7 station for a week.

Q *Name any of the non-American, non-Soviets who have flown in space through May 1984.*

A Vladimir Remek, Czechoslovakia; Miroslav Hermaszewski, Poland; Sigmund Jähn, East Germany; Georgi Ivanov, Bulgaria; Bertalan Farkash, Hungary; Pham Tuan, Vietnam; Arnaldo Tamayo-Mendez, Cuba; Jugderdemidin Gurragcha, Mongolia; Dumitru Prunariu, Romania; Jean-Loup Chrétien, France; Ulf Merbold, West Germany; Rakesh Sharma, India; Marc Garneau, Canada.

Q *Name the number and the crew of the final Gemini flight.*

A Gemini 12, November 11–15, 1966, with Jim Lovell and Buzz Aldrin. During the flight, his second Gemini mission, Lovell became the human with the most miles in space, reaching the fantastic total of 7 million.

Q *Who flew in the final Project Mercury flight?*
A L. Gordon Cooper, aboard *Faith 7*, May 15, 1963.

Q *Who was the first American scientist to go into space?*

A Depends how tightly you define "scientist." Buzz Aldrin (Gemini 12) had a Ph.D. in astronautics; Walt Cunningham (Apollo 7) was a research scientist for Rand Corporation when he joined NASA. But the nod generally goes to the first astronaut to go into space *as a scientist,* geologist Dr. Harrison Schmitt on Apollo 17, the last lunar mission.

Q *When did the Soviets last claim they would beat us to the Moon?*

A On October 23, 1968, in the person of Gherman Titov, during an interview in Mexico City. Titov had said earlier that the Soviets were also thinking of a circumlunar test flight, as we did with Apollo 8. A comment by Alexei Leonov in June 1969 that the Soviets would land on the Moon and retrieve samples in time for Expo '70 is considered a reference to Luna 16's automated retrieval.

Q *How many spectators were on hand for the launch of Apollo 11?*

A Just over 1 million, of which 3,497 were accredited journalists. Jules Verne would have been disappointed; he predicted a crowd of 5 million. The official emblem worn by the crew is shown here. Though the U.S. government restricted reproductions in any form (other than for news or education), it and similar versions became the most sought-after items at the Cape in the summer of 1969.

> **One Point of View** When Mike Collins did the preliminary sketch for the Apollo 11 mission patch, he put the wrong side of the Earth in shadow.

Q *Was there any significance to the order in which cosmonauts from "friendly" nations were chosen to fly in the Intercosmos program?*

A Rookies from the technologically advanced satellites of Czechoslovakia, East Germany, and Poland went first, followed by Bulgaria, Hungary, Vietnam, Cuba, Mongolia, and Romania. After much bewilderment, Kremlinologists realized it was simply Russian alphabetical order.

Q *Did Congress consider giving the Medal of Honor to the Apollo 11 crew?*

A Texas congressman Olin Teague wanted to introduce that resolution in 1969, citing Lindbergh's award. After agonizing over it, Deke Slayton told Teague it would be a slight to the other astronauts who had run interference for Apollo 11, and to the 410,000 workers who had been part of the mission. In fact, the crew had deleted their names from the mission badge for that reason. The Congressional *Space* Medal of Honor was established the same year, and later went to Neil Armstrong.

Q *Which astronauts have won the Congressional Space Medal of Honor?*

A The award, authorized in 1969, has gone to: Alan Shepard, Gus Grissom (posthumously), John Glenn, Pete Conrad, Frank Borman, Neil Armstrong, and John Young.

Q *What unexpected piece of equipment saved Armstrong and Aldrin from being stranded on the Moon?*

A A ballpoint pen. When the Moonwalkers climbed back into the LM, one of their life-support backpacks bumped the arming switch for the ascent stage, breaking it off. Houston advised them to retract the point of the pen and use the hollow end to reach the metal strip inside the broken switch. It worked. Little did Paul Fisher realize when he paid $1 million to develop a zero-g pen that it would become a lifesaver.

Q *Guess how much fuel remained when Apollo 11's LM landed on the Moon. None? Enough for 20 minutes? 2 hours? 20 seconds?*

A Just 20 seconds' worth.

Q *When did the Russians concede the Moon race?*

A In October 1969, Soviet Academy of Sciences president Mstislav Keldysh said in Stockholm, "At the moment, we are concentrating wholly on the creation of large satellite stations. We no longer have any scheduled plans for manned lunar flights."

Q *Identify the only U.S. president to personally come aboard a recovery ship to greet the returning crew of a spaceflight.*

A Richard M. Nixon, who was aboard the USS *Hornet* when it recovered Neil Armstrong, Buzz Aldrin, and Michael Collins after the first lunar landing, Apollo 11, in July 1969.

Q *Why wasn't President Nixon allowed to have dinner with the astronauts the evening before Apollo 11's lift-off?*

A The crew was in prelaunch quarantine, and it was feared that the President or other outsiders might carry stray bacteria that would attack the crew during the lunar mission. Lyndon Johnson was similarly barred, even though both men had been invited by the crew. In this photo Nixon talks to the Apollo 11 Moonmen via a very-long-distance phone call.

Q *How long did the Apollo 11 astronauts remain in quarantine after returning to Earth from the first manned landing on the lunar surface?*

A Neil Armstrong, Michael Collins, and Buzz Aldrin were housed in a special quarantine van for eighteen days after returning to Earth. At this time it was not known if any possibly contaminating "lunar germs" could have returned with them. President Nixon, seen with the astronauts aboard the USS *Hornet,* talks to them via microphone.

Q *Name the last crew to be quarantined after returning from the Moon.*
A Apollo 14 (Shepard, Roosa, Mitchell). Missions 11, 12, and 14 failed to bring back any intergalactic plague, so the practice was deemed unnecessary. This photo shows Shepard, the first American in space, deploying the U.S. flag during the Apollo 14 mission.

> **"Dam" Powerful** During each Project Apollo launch, the Saturn rocket created eighty-five times as much power as does Hoover Dam.

Q *What familiar symbol appeared on Project Apollo's Saturn rockets, and where were the symbols made?*

A The American flag, silk-screened onto adhesive paper and pasted onto the rockets as decals. They were made at the Boeing Aircraft plant in Michoud, Louisiana.

Q *When was the first American flag worn into space?*

A The Stars and Stripes had been painted on every spacecraft, but it wasn't until Gemini 4 that astronauts wore the flag on their suits. McDivitt and White were denied permission to name their craft *American Eagle.*

Q *How did the Wright Brothers get to the Moon?*

A Armstrong and Aldrin carried a piece of the first airplane's fabric and propellor. The *Kitty Hawk* was built and flown in 1903.

Q *What is armalcolite?*

A A new mineral found by the Apollo 11 crew at Tranquility Base on the Moon. The name is a contraction of the names of the crew. The mineral is also known as $(Fe^{2+}, Mg)Ti_2O_5$.

Q *Which of the Original 7 Astronauts tried to convince NASA officials that one more flight should be added to the Project Mercury program?*

A Alan Shepard, who wanted to pilot a three-day mission. He was supported by several of the other astronauts, but President Kennedy let it be known that Project Mercury was finished after Gordon Cooper's *Faith 7* flight: the space race would now be moving on to Project Gemini.

Q *Who was the only astronaut to lose his spacecraft?*

A Gus Grissom. *Liberty Bell 7*'s hatch blew prematurely, and the craft took on water and sank. While some privately blamed Grissom for bumping a switch, an inquiry showed that had he actually done that, there would have been telltale abrasions or bruises on his arm. NASA ruled it a malfunction.

Q *When was the first Apollo mission originally slated to fly?*
A Grissom, White, and Chaffee (bottom to top) were to have flown a joint mission with Gemini 12 (Lovell, Aldrin) in November 1966. Construction delays made that impossible. This photo was taken on January 25, 1967, just two days before their death.

Q *Name the three of the Original 7 Astronauts who went into space during the Gemini program.*

A Gus Grissom was aboard the first manned Gemini mission, launched March 23, 1965 (with John Young). Gordon Cooper joined Pete Conrad in the third Gemini mission, on August 21, 1965. And Wally Schirra was on the fourth Gemini flight, launched December 15, 1965 (with Thomas Stafford).

Q *Only one of the Original 7 Mercury Astronauts went into space on Mercury, Gemini, and Apollo missions. Name him.*

A Walter Schirra, who was solo aboard *Sigma 7*, lifted off with Thomas Stafford in Gemini 6 (the fourth manned flight in the series), and was part of the first crew in the three-manned Apollo series aboard Apollo 7, when his crewmates were Donn Eisele and Walter Cunningham. In the photo here, Schirra trains for his Gemini mission.

Q *Did anyone actually see the Apollo 13 explosion?*

A Yes. A team of engineers at Johnson Space Center, headed by Andy Sauletis, had attached a TV camera to a telescope atop one of the buildings in Houston. They were tracking the spacecraft visually and spotted a widening circle of light where 13 should have been. They thought it was a malfunction in their monitor, but it turned out to be the liquid oxygen escaping from the ruptured tanks. This dramatic photo, taken by the crew from the lunar module, shows the area where the panel was blown away from the service module. The Apollo 13 crew had to use the lunar module as a "lifeboat" until it was jettisoned prior to Earth re-entry.

> **Who's Afraid of 13?** The ill-fated flight of Apollo 13 gave triskai-dekaphobia some credence:
> • The explosion took place on April 13.
> • Launch time was 13:13, Houston time.
> • The crew had considered a black cat insignia.
> • There are 13 letters in "German measles," the disease that grounded Ken Mattingly.
> • There are 13 letters in the names "James, Fred, Jack," the astronauts who finally flew.
> The "Zero Year" theorists have plenty to talk about, too: Apollo 13 was the first mission of 1970, which means the whole thing will happen again in twenty years.

Q *Which spacecraft was struck by lightning at launch?*

A Apollo 12 was struck, not once but twice, on its ascent into overcast skies, November 14, 1969. The lightning temporarily knocked out onboard electronics.

Q *What seismic activity was recorded by an experimental package left by Apollo 12?*

A The impact of Apollo 13's third-stage Saturn 4-B booster on the lunar surface. The equipment in the Ocean of Storms registered shock waves for over 4 hours.

Q *Which astronauts described air on the Moon?*

A Pete Conrad and Alan Bean of Apollo 12. They spent a few minutes tossing a canister lid from an experimental package back and forth like a Frisbee, then commented on how well the "lunar air" held up the saucer. There is no air on the Moon; the decreased gravity made the lid behave so well.

Q *Which was the first Apollo mission to have an all–Air Force crew?*

A Apollo 15: Dave Scott, Jim Irwin, and Al Worden. A sign awaited them on launch day: THIS FLIGHT IS BEING BROUGHT TO YOU THROUGH THE COURTESY OF THE NAVY.

Q *Whose idea was it to read from Genesis on Apollo 8?*

A The idea came from Si Bourgin of the U.S. Information Agency. It took quite a bit of selling for Frank Borman to get Catholic Bill Anders, not to mention nonreligious Jim Lovell, to read from the King James version.

Q. *Which one of the nine astronauts selected in Group 2 was not a member of the armed forces or already employed by NASA?*

A Elliot See, who was a test pilot for General Electric. The only other civilian in the group was Neil Armstrong, a test pilot with NASA. See is seen above entering a F-104 jet.

> **Where Did the Day Go?** During orbital flights the Project Mercury astronauts saw sunsets at a rate eighteen times faster than seen on Earth.

Q *How did Thomas Stafford and Eugene Cernan become the prime crew for Gemini 9?*

A When Charles Bassett and Elliot See, the original prime crew, were killed in a plane crash on February 28, 1966, Stafford and Cernan, the backup crew, took over. They lifted off on June 3, 1966.

Q *Identify the first NASA astronaut or trainee to lose his life.*

A Theodore Freeman, who was killed in the crash of a T-38 jet at Ellington Air Force Base, Houston, on October 31, 1964, after a goose crashed into his cockpit.

Q *Name the first man to die in spaceflight.*

A Vladimir Komarov, during the first flight of the Soyuz spacecraft, on April 24, 1967. During re-entry the parachute deployed automatically, but it tangled around the Soyuz as it tumbled end-over-end. Komarov hit the ground at about 400 mph.

Q *When was the first assassination attempt made on a space traveler?*

A On January 22, 1969. As the Soyuz 4 and 5 crews were returning to Red Square from Vnukovo Airport, an army officer opened fire with a machine gun on one of the limousines. The driver was killed, a motorcycle cop was wounded, and Col. Georgi Beregovoy was injured by flying glass. Also in the car were Alexei Leonov, Andrian Nikolayev, and Valentina Tereshkova. The attacker thought the car was carrying Secretary Leonid Brezhnev and President Nikolai Podgorny. Nikolayev bears a resemblance to Brezhnev.

Q *Name the first mission on which frogmen were dropped into the water to attach flotation and assist in recovery.*

A MA-6, John Glenn's flight. After Grissom's near-drowning, the spacecraft also carried a life raft.

Q *Which astronaut helped develop the Sidewinder missile?*

A Wally Schirra, during a stint at China Lake, California. He also worked on the F7U3 Cutlass fighter.

Q *Who showed dissatisfaction with a spacecraft by hanging a lemon on it?*

A Gus Grissom hung a lemon on the Apollo 1 command module the week before he died in it. He had complained loudly within NASA about the 20,000 test failures logged in checkout of the cabin and engines.

Q *What was Betty Grissom's response to learning that Gus had won the Distinguished Flying Cross posthumously?*

A "Medals don't mean much to me." She declined an invitation to the White House to accept it. However, in 1978 Mrs. Grissom did accept the Congressional Space Medal of Honor for Gus, which President Carter presented at the Kennedy Space Center.

Q *Where are the Apollo 1 astronauts buried?*

A Gus Grissom and Roger Chaffee are buried at Arlington National Cemetery; Ed White is buried at his alma mater, West Point. Here fellow astronauts Alan Shepard (left), John Glenn, Gordon Cooper, and John Young escort the flag-draped coffin of Gus Grissom into Arlington. Not visible in the photo are Scott Carpenter, Deke Slayton, and Wally Schirra.

Q *Has there ever been a launchpad explosion of a booster vehicle with a crew aboard?*

A Yes. A Soviet A-2 booster exploded during the launch of a Soyuz spacecraft on September 27, 1983. The launch tower separated the Soyuz from the booster just before the explosion, dropping the cosmonauts Vladimir Titov and Gennadi Strekalov nearby. They reportedly sustained injuries, but survived. The Soyuz is the first Soviet spacecraft to have such an escape system.

Q *What caused the fatal fire aboard Apollo 1?*

A According to the investigation board's findings, the fire was started by a short circuit in the wiring beneath Grissom's couch. Among the contributing factors cited: inadequate provisions for crew escape, rescue, and medical assistance; a sealed hatch that took five minutes to open; a cabin full of pure oxygen at high pressure; no onboard fire extinguisher; external fire extinguishers that were empty or underpressured; pad worker masks made for poison gas, not smoke; shoddy wiring and plumbing; and the presence of combustible materials in the cabin. Those who testified at the congressional hearings held by the House Space Committee in 1967 included, from the left, astronauts Frank Borman, Jim McDivitt, Deke Slayton, Wally Schirra, and Alan Shepard.

Got Bad News Together—When the news of the fatal Apollo 1 fire reached the White House, President Lyndon Johnson was able to talk to the key NASA administrative personnel face to face. Earlier in the day the Space for Peace treaty had been signed there, and in attendance were: NASA administrator James Webb, JSC director Robert Gilruth, KSC director Kurt Debus, and MSFC director Wernher von Braun. The photo above was taken shortly after the accident.

> **Cigarette Saved His Life** A craving for tobacco saved the life of Russian space official Mikhail Yangel during the 1960 Tyuratam explosion. After the rocket failed to ignite, he was one of those inspecting the rocket on the pad. When he had a sudden urge for a cigarette, Yangel climbed down into a blockhouse as safety rules demanded. No sooner had he lit up than the rocket exploded, killing almost everyone else on the pad.

Q *What was the Tyuratam Slaughter?*

A Also known as the Nedelin Disaster, it was the explosion of a large booster on the launchpad in October 1960. When the rocket refused to ignite, officials waived normal safety constraints and moved an inspection team onto the pad. Suddenly the rocket ignited and exploded. Estimates of the number of dead range from 40 to 300, including Russia's top military missile authority, Field Marshall Mitrofan Nedelin.

Q *How hot was the Apollo 1 fire?*

A The interior of the spacecraft reached 1,000° F.

Q *Which Soyuz mission crashed into a mountain range?*

A An unnumbered Soyuz on April 5, 1975, known simply as the "April 5th Anomaly" or "Soyuz X." The craft failed to separate from the last stage of the booster, dragging it back down to Earth. Vasily Lazarev and Oleg Makarov managed a crash landing into a snowbank in the Altai mountain range. After touchdown, they began sliding toward a steep drop and were saved from the chasm only when their parachute lines snagged around some scrub pines. Lazarev sustained serious injuries and never flew again.

Q *Identify the souvenirs that Gus Grissom carried with him during his* Liberty Bell 7 *flight.*

A Gus had wanted to take one hundred dollars in singles, but since he didn't have that much money to spare, he settled for two rolls of dimes, three one-dollar bills, and several small models of his spacecraft.

Q *How much did Betty Grissom receive in her lawsuit following Gus's death?*

A North American Rockwell settled out of court in a pain and suffering claim, giving Grissom's family a total of $350,000. Although they weren't part of the original suit, the families of Ed White and Roger Chaffee received the same amount.

Q *How much weight did John Glenn lose during his flight?*

A Between the time of launch and the examination Glenn received aboard the USS *Randolph*, he lost 5⁵⁄₁₆ pounds. It was approximately the same weight loss he had experienced during simulated tests on the centrifuge.

Q *If the Vehicle Assembly Building were divided into apartments, how many people could live there?*

A According to engineer Jim Griever, who ran a computer analysis of the question in 1979, the 129,000,000-cubic-foot building could be divided into 9,000 apartments to house 27,000 residents. The question arose when co-author Brian Wilkes had trouble finding an apartment in housing-short Melbourne.

But How Big Is It?   The Vehicle Assembly Building at Kennedy Space Center is the second-largest building (by volume) in the world. The United Nations Secretariat Building could fit through its door. There is room on the roof for the Houston Astrodome and most of its parking lot, and room inside for 3¾ Empire State Buildings.

> **If Looks Could Kill** An unverified story claims that at a mid-1960s news conference, a reporter asked Wernher von Braun if there was any chance that the experimental Saturn 5 might go off course and strike London. Von Braun allegedly glared at the man and walked out. NASA was, and still is, touchy about the former political affiliations of some of its employees.

Q *What was the only signal failure on board the spacecraft during Alan Shepard's* Freedom 7 *flight?*

A The green light on the instrument panel, which was supposed to indicate that the retro-package had been successfully jettisoned, failed to work. However, Shepard had a visual sighting of the pack as it fell toward the ocean.

Q *Who was John Glenn's backup pilot for the first orbital mission?*

A Scott Carpenter. Alan Shepard served as the technical advisor.

Q *Which Project Mercury flight was the only one to return to Earth with manually controlled retro-fire and re-entry?*

A Gordon Cooper's *Faith 7* (MA-9), the last flight in the program, May 15, 1963.

Q *Who was the first human to change the direction of a spacecraft's orbital path?*

A Gus Grissom, during a test in the flight of Gemini 3, the *Molly Brown*, on March 23, 1965. Co-crewmember was John Young.

Q *Did Gagarin manually control his spacecraft?*

A No. The control console was locked up, to prevent him from touching the controls unless an emergency arose, when he would have been radioed the combination (1-4-5).

Q *Name the first Soviet cosmonaut to take manual control of his spacecraft.*

A Pavel Belyayev, on Voskhod 2. The automatic landing system failed, forcing Belyayev and Leonov to perform the landing themselves. They missed the central Asian landing zone by over 2,000 miles, landing instead in the northern Urals, where they spent a freezing night convincing wolves that they were not a canned meal.

Big Apple "Juice" The first two stages of the Saturn 5 generated enough energy to supply electricity to all of New York City for 1¼ hours.

Who Was That Lady? Media and public interest in the lives of the Project Mercury astronauts got so out of hand for a time that when Rene Carpenter demanded privacy and told some members of the press not to try to locate her during her husband's flight, it was as if she had declared war. So she resorted to deception. In order to reach Cape Canaveral unnoticed, she had her four children lie on the floor of her car, since the TV networks had people checking the bridges and causeways for a woman with children. There were also helicopters circling above the traffic, specifically looking for her car.

Q *Name the first reusable spacecraft booster.*

A Most people would say the space shuttle solid rocket boosters, but the correct answer is the Vostok booster, also known to American observers as the A-1. The Russians built the rocket so heavy that parts of it were often recovered downrange in Siberia and reused.

Q *Name the first spacecraft to fly more than one mission.*

A The craft that we know most commonly as Vostok 1. According to the Soviets, the vehicle flew two missions before it took Yuri Gagarin on man's first orbital flight.

Q *Name the primary Soviet launch site.*

A All of the manned launches to date have left from Star City, a base in the semi-desert of southern Kazakhstan. It was opened in 1957 as the Baikonur Cosmodrome; Baikonur is actually over 200 miles away, but Americans saw through the ruse and refer to it as Tyuratam, after the closest town. The new city that has grown there is called Leninsk. Take your pick!

Q *Who owned the land that became Johnson Space Center?*

A Some was owned by Humble Oil & Refining, some by Rice University. A clause in the purchase agreement states that the buildings must be designed to be easily convertible to use by Rice if NASA ever abandons the land, which explains why much of it already looks like a college campus.

Q *Which mission marked the first use of the Mission Control Center in Houston?*

A Gemini 4. On previous flights, launch and control operations had been carried out in Florida, with monitoring at Goddard Space Flight Center, in Greenbelt, Maryland.

Q *Name the first astronauts to orbit the Earth while awaiting a court appearance.*

A Wally Schirra and Walt Cunningham. Just before their Apollo 7 mission, the pair was busted for drag racing in Cocoa Beach. Try as he might, Wally couldn't charm the officer into dropping the charge. The charge was dropped by a county judge after the mission.

Q *When was artificial gravity first created in space?*

A During the Gemini 11 flight, in 1966. The spacecraft was tethered to an Agena target vehicle by a long Dacron line, causing the two vehicles to spin slowly around each other. According to Pete Conrad, a TV camera fell "down" in the direction of the centrifugal force.

Q *Identify the designer of the lunar module.*

A John C. Houbolt. He designed it in 1961, eight years before one would actually land on the surface of the Moon. Houbolt received approval for his design in 1962 despite objections from people such as Wernher von Braun who favored using two spacecraft (one as a fuel tanker and the other to blast off to the Moon from space).

Mileage Check   During the Project Mercury orbital flights, the exact distance from the moment the retro-rockets fired for re-entry to the point of splashdown in the ocean was 2,990 miles.

Q *Name the former Indianapolis 500 winner who operated a General Motors automobile dealership near Cape Canaveral and supplied the Original 7 with fast cars.*

A Jim Rathman, who had won at Indy in 1960 with an average speed of 138.767 mph.

Q *Which astronaut first wore red racing stripes on his spacesuit during a Moonwalk?*

A Alan Shepard, on Apollo 14. All subsequent lunar mission commanders did likewise.

Q *Which astronaut hit a golf ball on the Moon?*

A Al Shepard, on Apollo 14. Though it took him three swings to connect with the ball, he claims he made a hole-in-one—into a crater.

Q *How many delays took place before John Glenn made the historic orbital flight in* Friendship 7 *?*

A Six times, because of bad weather and technical difficulties.

Q *Why was Alan Shepard's flight delayed from March 12 to May 5?*

A Perfectionist Wernher von Braun insisted on one more test-firing of the Redstone rocket. Had Shepard flown on March 12, he would have beaten Yuri Gagarin into space by exactly one month.

Q *Identify the first two astronauts to transfer from one spacecraft to another.*

A Jim McDivitt and Rusty Schweickart, in Apollo 9, when they moved from their command module (*Gumdrop*) into the lunar excursion module (*Spider*) to test it in Earth orbit. The launch date was March 3, 1969.

Q *Identify the legendary performer who entertained the Original 7 during the Fourth of July party given for them in Houston in 1962.*

A Sally Rand, the lady who made the fan dance famous when the Original 7 were in their teens.

Q *What was America's first orbital emergency?*

A Discounting the re-entry problems of Glenn, Carpenter, and Cooper, the first near-disaster was on Gemini 8. After Neil Armstrong and Dave Scott docked with the Agena target vehicle, a runaway thruster caused both craft to spin wildly. When they separated the two vessels, the spinning increased to almost 60 rpm. Bringing it under control, they landed at the next opportunity, cutting the mission short.

Q *Were Armstrong and Scott ever close to blackout from the rapid spinning of Gemini 8?*

A Not really, according to Armstrong, who reminds us that conventional blackout is unlikely in weightlessness. He does admit that disorientation and the inability to operate switches and controls was a major concern.

Q *Identify the very appropriate song recorded by Frank Sinatra, and the Apollo crew that played it.*

A "Fly Me to the Moon" was played by the crew of Apollo 10—Tom Stafford, John Young, and Gene Cernan—during their mission to test the LM in lunar orbit. They came to within 10 miles of the Moon's surface, and the following morning Ground Control awakened them by playing a tape of Robert Goulet singing "On a Clear Day You Can See Forever." Stafford and Young, fourth and fifth from left, are seen here in training.

> **Earthbound Copycat** The National Air and Space Museum in Washington, D.C., houses a full-scale model of Skylab. A few hours before the real thing crashed, a large piece fell off the model.

Q *What song did Scott and Irwin play as they left the Moon?*

A Mission Control was treated to "Wild Blue Yonder," in honor of the first all–Air Force Crew.

Q *John Glenn played the record of an opera so frequently prior to his flight that he feared he would wear it out. What was it?*

A Puccini's *Madame Butterfly.*

Q *Who suggested that Muzak be installed on spacecraft for long-duration flights?*

A Gordon Cooper and Pete Conrad made the suggestion as a joke, after music was piped up from Houston.

Q *What song did Yuri Gagarin sing in space?*

A During the descent, he sang the folksong "The Homeland Hears, The Homeland Knows."

Q *When did Skylab crash?*

A On Wednesday, July 11, 1979, near the town of Esperance, in western Australia. The only confirmed casualty was a jackrabbit, found in the crater made by the impact of an oxygen tank.

Q *With what crime did local officials of Esperance, Australia, charge NASA after the Skylab crash?*

A Littering. When a team from Huntsville arrived to pick up the pieces, they were handed a littering citation and asked to pay a $400 fine. The Aussies played it so straight the Yanks weren't sure if it was a gag or not.

Q *On which mission did the first strike in space occur?*

A On the Skylab 4 mission, astronauts Pogue, Carr, and Gibson became disgusted with their work load and ignored orders from Houston. They spent a day taking pictures and doing experiments that were interesting to them. Similar shutdowns have taken place on Soviet Salyut missions.

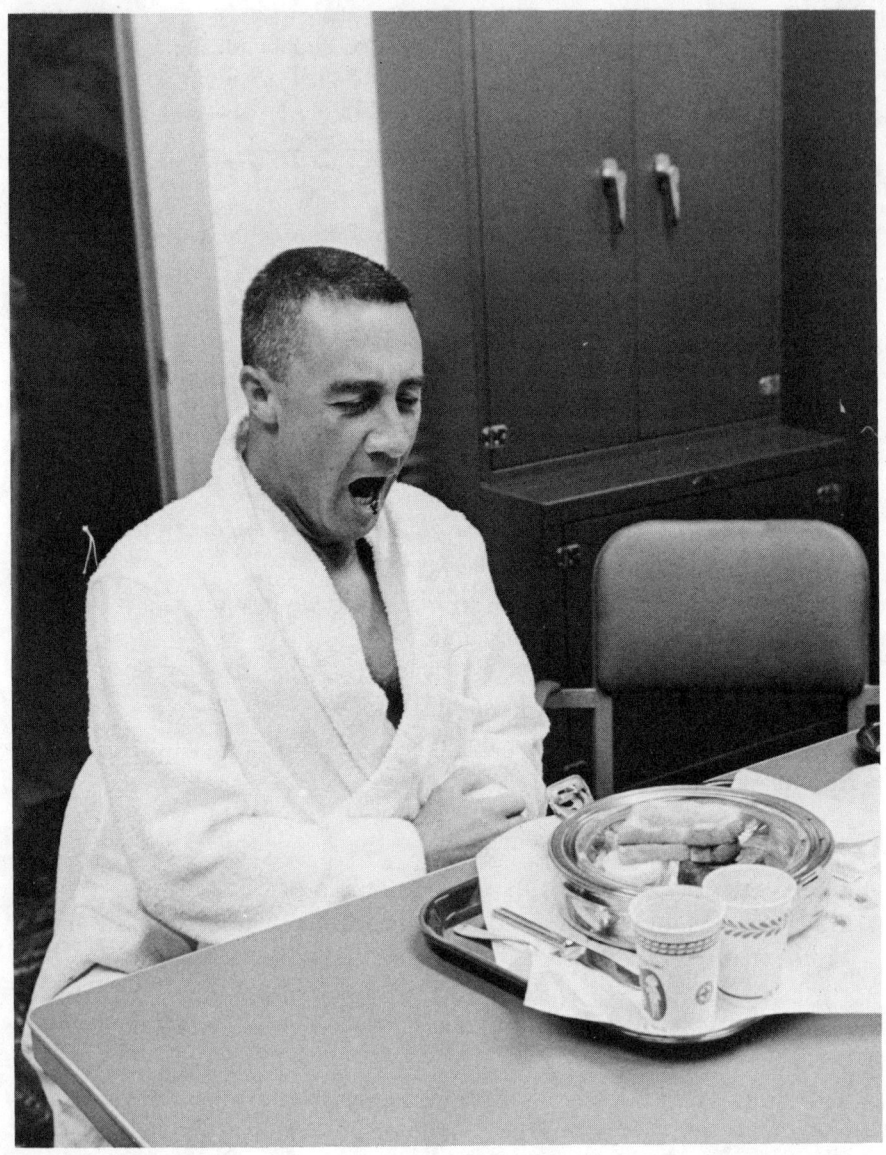

Q *Who started the custom of the prelaunch breakfast of steak and eggs?*

A Alan Shepard, on the first flight. While there's no dietary imperative for that fare, few of the astronauts have ever broken tradition by ordering something else. As NASA spokesman Mark Hess put it, "Given the choice, any *real* American wants steak and eggs." However, Gus Grissom seemed less than enthusiastic before starting his 2:15 A.M. breakfast in Hanger S.

> **Where's the Beef?**   Before flying the shuttle *Enterprise* on approach and landing tests in California, Fred Haise ordered the traditional steak and egg breakfast. The cook brought him corn flakes, toast, and prune juice, reminding him that it wasn't a "real" space flight.

Q *Who was the first astronaut to smuggle food into space?*

A John Young. His Gemini 3 companion, Gus Grissom, loathed the packaged "spacefood." Young had two sandwiches made at a Cocoa Beach restaurant, and into the first revolution, he unzipped a leg pocket, produced the contraband, and said "Care for a corned beef sandwich, Skipper?" Deke Slayton says Wally Schirra was behind the whole thing.

Q *Identify the first astronaut to have the opportunity to eat a steak sandwich in space.*

A Wally Schirra. The sandwich, and a gag ignition key mounted on his right-hand attitude controller, were placed in *Sigma 7* by pad technicians.

Q *Identify the astronaut who won a steak dinner from Alan Shepard as a result of seeing a star in space.*

A Gus Grissom. Shepard had bet him that he wouldn't, even though Grissom's Mercury spacecraft had a new 19-inch picture window (Shepard's craft had had only two portholes). As Grissom entered the darkness of space, he called back to Earth, "I see a star." Shepard, his capsule communicator, then owed him a meal.

Q *What alternative source of water for the astronauts was available for the first time, and on which mission?*

A Water generated by the fuel cells was available starting with Gemini 3, but the water wasn't drunk because of a high pH count.

Q *Name any items in the first meal eaten on the Moon.*

A Unlike Jules Verne's cosmic gourmets, Armstrong and Aldrin had bacon squares, sugar cookies, canned peaches, pineapple-grapefruit juice, and coffee.

Q *Who was the first administrator of NASA?*

A T. Keith Glennan, president of the Case Institute of Technology.

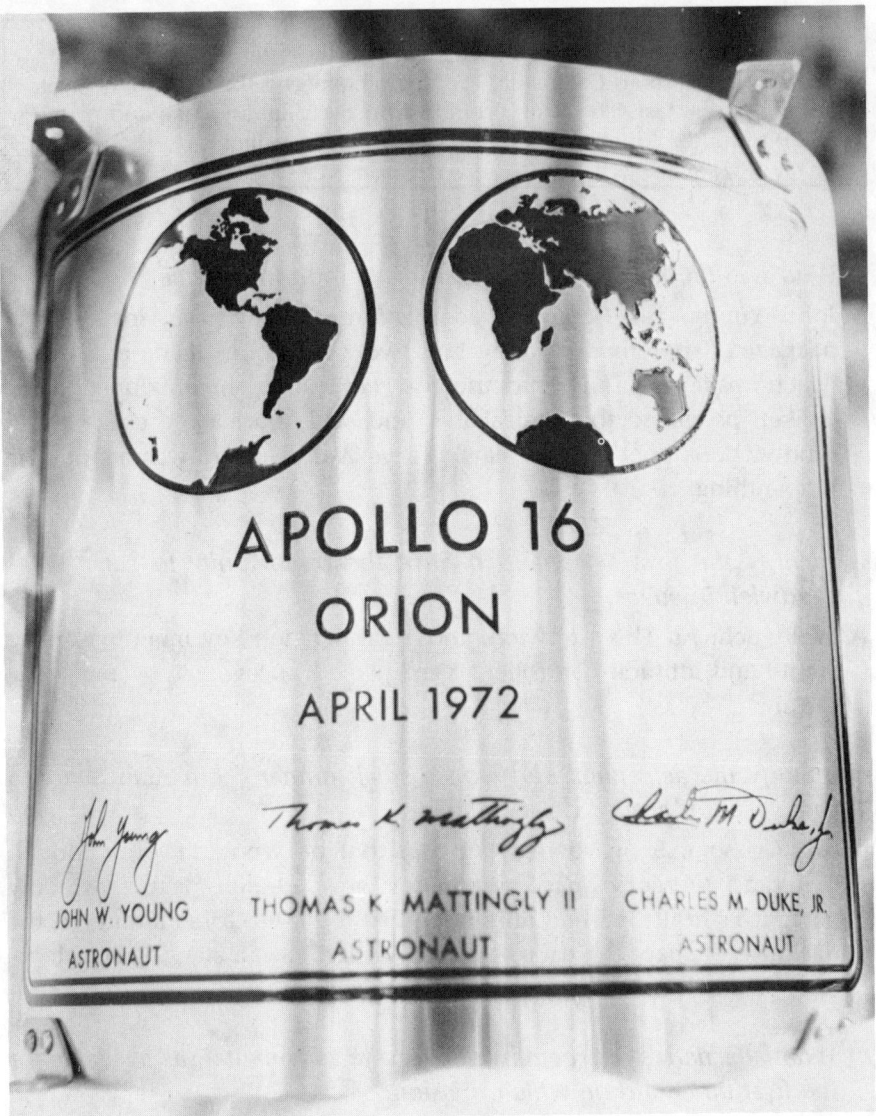

Q *Which astronaut made "colorful" complaints from the Moon about the menu?*

A Floridian John Young, from the citrus-belt city of Orlando, blamed his Apollo 16 flatulence on orange juice. Not realizing that his helmet mike was relaying his comments back to Houston, he swore to Charles Duke: "It's only twelve fucking days, I ain't ever going to eat them anymore. If they offer me some to eat . . . I'm going to throw up." The photo above is of the plaque left by the Apollo 16 crew on the Moon.

> **And Don't Drink the Water** On forms evaluating the palatability of various spacefood recipes, Francophile Mike Collins drew crossed knives and forks and small stars, in the manner of the *Guide Michelin*, to indicate his rating. NASA was not amused.

Q *What new spacefood was introduced on Gemini 5?*

A A synthetic orange juice called Tang. It has remained on the menu ever since.

Q *Who was the first American to vomit in space?*

A Frank Borman, on Apollo 8. The incident of spacesickness occurred on the way to the Moon, making him also the first man to vomit beyond Earth orbit. It's believed that Americans weren't affected as early in the program as the Russians because of the relatively smaller size of the spacecraft. Borman's illness was kept from the media at the time, while NASA officials tried to decide whether it was serious enough to order a scrub.

Q *Did NASA ever include alcoholic beverages on the space menus?*

A Before the Skylab series began, NASA thought it wouldn't hurt to include some wine with dinners, and went as far as holding a wine-tasting party in Houston to learn the astronauts' preferences. Temperance groups heard of the plan and deluged the agency with protest letters, so the crew of Skylab had to make do with Tang.

Q *With what simple treats did Viktor Patsayev celebrate his thirty-eighth birthday while in space?*

A Although he came from a country known for caviar, Patsayev marked the day in June 1971 with a lemon and an onion, two pieces of "real" food smuggled from Earth to Salyut. He told Andrian Nikolayev that he had been unable to smuggle any vodka aboard.

Q *Who took the first liquor into space?*

A The first confirmed bottle, a miniature of Scotch, was reportedly aboard Wally Schirra's *Sigma 7* Project Mercury flight. Miniatures of brandy were also smuggled aboard the first lunar flyby, Apollo 8, so the crew could celebrate Christmas.

Q *Who were the first humans to see the far side of the Moon?*

A Frank Borman, Jim Lovell, and Bill Anders, during the flight of Apollo 8, December 21–28, 1968.

Q *After John Glenn's flight, the Original 7 Astronauts and their wives were honored with a ticker-tape parade in New York City. Name the musical they went to see that night.*

A *How to Succeed in Business Without Really Trying.* The decision to go was made during a dinner being given for them all by Henry Luce, of Time, Inc. The start of the play was delayed while the management arranged for the necessary number of seats to be grouped together. Theatergoers gave up their seats with little hesitation.

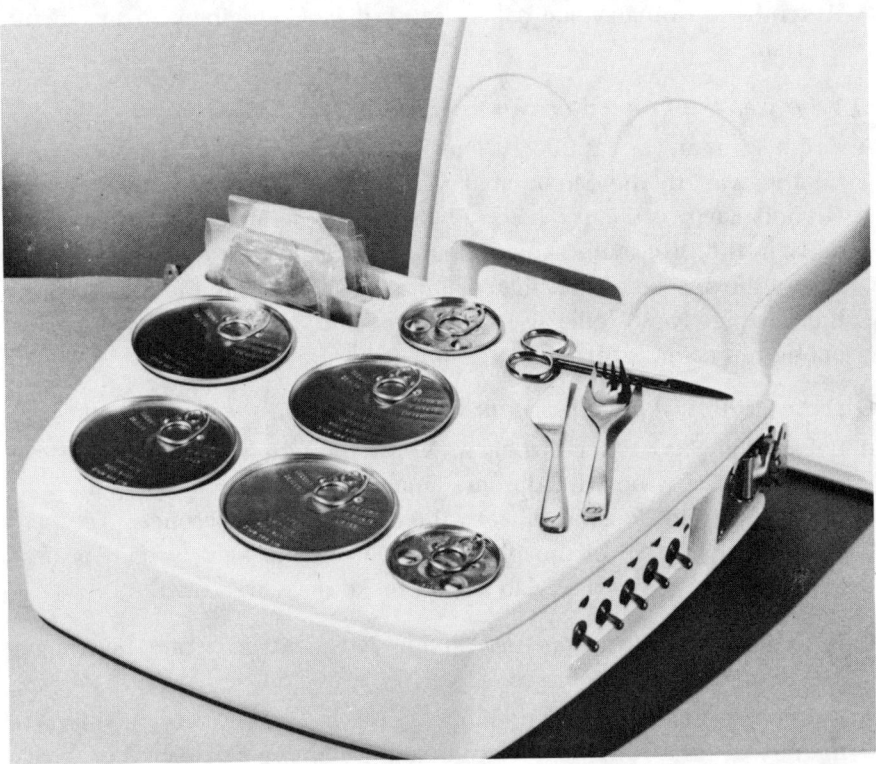

Is This All On One Check?—Three meals were prepared for each eating period in the three days before a Project Mercury flight. The designated astronaut ate one, the backup pilot another, and the third was frozen. The purpose was to determine if any specific food might cause problems. By having the backup pilot consume the same food as the primary pilot and also having another portion frozen, NASA hoped to identify any problems. By the time Skylab became operational, the food astronauts ate both on Earth and in space had come a long way. The tray above was typical fare: the crew had a choice of various frozen foods and conventional spacefood as used on earlier missions. Note the switches on the side of the trays for heating canned foods.

> **Clothes Make the Man**   The spacesuits worn by Wally Schirra, Walt Cunningham, and Donn Eisele for the first manned Apollo mission cost $100,000 each. The suits weighed 57 pounds apiece, were individually tailored, and contained twenty-one layers of fireproof fiberglass and aluminized plastic. They were designed to protect the astronauts from temperatures that range from −250° F to +230° F.

Q *Name the first astronauts to experience head colds during a mission.*

A Wally Schirra, Donn Eisele, and Walt Cunningham, all during Apollo 7, October 11–22, 1968.

Q *Identify the only two astronauts/cosmonauts to appear on covers of* Time *magazine in 1961.*

A Cosmonaut Yuri Gagarin and astronaut Alan Shepard.

Q *Which magazine bought exclusive book and magazine rights for the Original 7 Astronauts' personal stories, and how much did it pay?*

A *Life* magazine paid $500,000 over three years, to be shared equally. The idea was that selling exclusive rights to a single news organization would protect the astronauts from requests for time-consuming interviews by other journalists.

Q *Who publishes the newspaper* Neptune?

A The Soviet cosmonauts. While much of the paper is fiction, poetry, and artwork, it also contains inside information and often scathing political cartoons. *Neptune* is considered a classified document, and at Soviet space installations it is posted on bulletin boards where all copies can be accounted for.

Q *How did John Young respond to an alarm light on his return to Earth?*

A An old test pilot's adage is that if a red light flashes, tap it with a hammer to see if it goes out. Young went one better—he kicked the control panel.

Q *Identify the crew of the U.S. spacecraft that successfully completed the first docking in space.*

A Neil Armstrong and David Scott, aboard Gemini 8, March 16–17, 1966. They docked with an Agena rocket.

Q *How many languages were spoken on the Apollo-Soyuz mission?*

A Officially there were only two, English and Russian, but the Soviets claim that Tom Stafford's heavily accented Russian comprised a separate language, "Oklahomski." And a fourth language, developed from joint slang, was called "Ruston" (Russian + Houston). Stafford and cosmonaut Alexei A. Leonov are seen here examining food for the historic flight during training at Johnson Space Center, Houston.

> **AAA Wouldn't Come** After the Grumman-built lunar module ferried the Rockwell-built command module back from the Moon, Grumman submitted a bill to Rockwell . . . for towing!

Q *Who played "knock-knock" in space?*

A Tom Stafford. After docking with the Soyuz, procedure called for Leonov to crawl through the adapter into the Apollo, and to knock on the hatchway for admission. When Leonov knocked, Stafford answered in Russian, "Who's there?"

Q *Did American companies serve as contractors to the Russian space program?*

A At least one did, the Fisher Pen Company of Van Nuys, California, which provided pressurized ballpoint pens to NASA. Cosmonaut Alexei Leonov urged the Soviets to order 100 pens and 1,000 ink cartridges, since Soviet pens had failed and cosmonauts used lead pencils, which were sometimes hard to read in reduced lighting.

Q *How many companies were involved in the U.S. space program between 1959 and 1969, when Apollo 11 landed on the Moon?*

A More than 20,000 companies were in some way involved. At the peak, during Project Apollo, over a half a million people were employed in jobs connected with the space race.

Q *Identify the U.S. company that produced the Mercury capsules, or spacecraft.*

A McDonnell-Douglas. They manufactured a total of twenty.

Q *Identify the U.S. politician who encouraged Soviet premier Nikita Khrushchev to have the U.S.S.R. join the U.S. in a joint effort to reach the Moon.*

A President John F. Kennedy, one month after he had made a congressional address setting a goal of landing a man on the Moon by the end of the decade. According to historians, almost immediately after Kennedy boldly announced the Moon landing goal, he regretted his impulsiveness, and he more than once wished he had never made the statement. However, JFK repeated the offer during a United Nations speech on September 20, 1963, despite high-level opposition within NASA and his own fears that the goal might be too costly and difficult to meet.

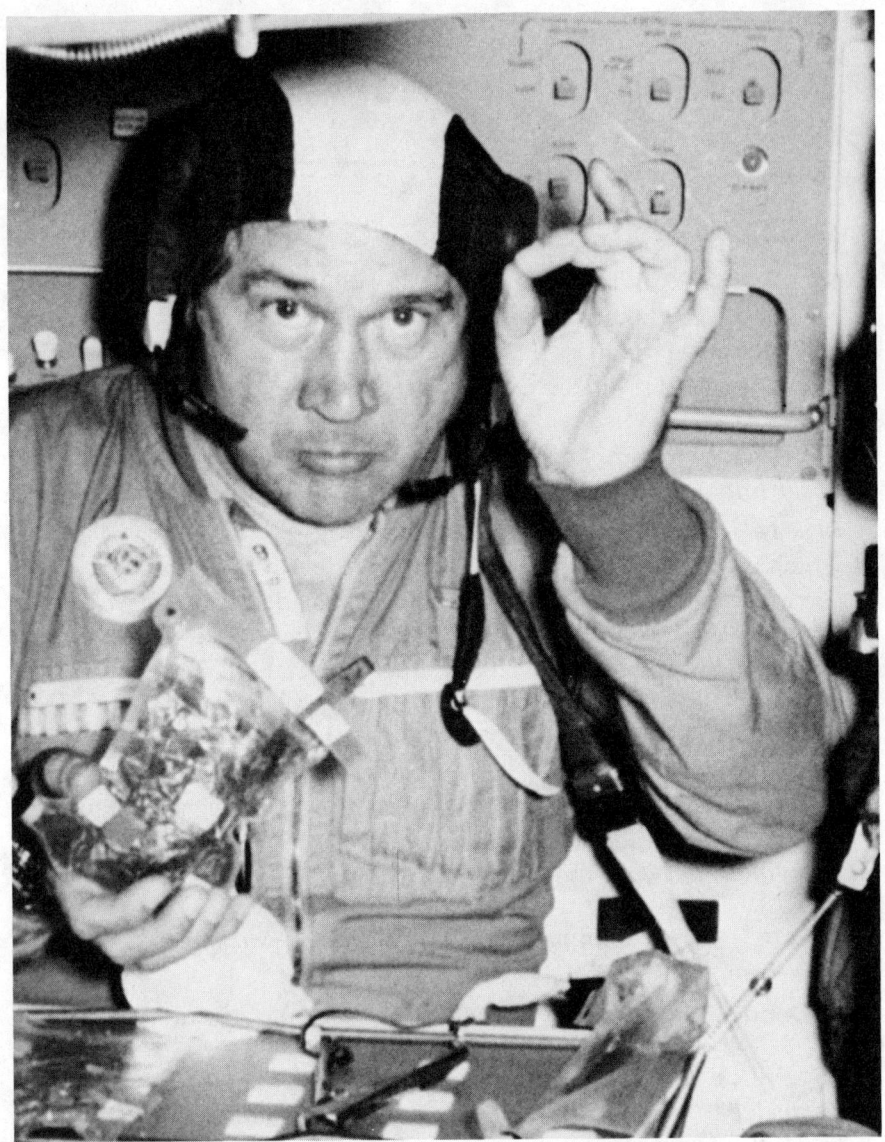

Q *What was the Strawberry Cube?*

A A processed dessert devised by Dr. Malcolm Smith, NASA's top nutritionist, in the early 1970s. It was considered almost perfect in terms of nutrition, design, shelf life, fragmentation in zero-g . . . but the nauts wouldn't eat the stuff. The cubes were later encased in Lucite and handed out to visiting VIPs. Here, Soviet cosmonaut Anatoly V. Filipchenko signals his approval of other American space cuisine during joint training for the Apollo-Soyuz mission.

Nicknames,
Code Names

Q *What is the name of the zone of radiation that creates a ribbon around the Earth?*

A The Van Allen Belt, named after Dr. James A. Van Allen, who was active in the development of the Explorer satellite and is credited with discovering the zone of radiation that carries his name.

Q *Who is known as "the father of modern rocketry"?*

A Robert H. Goddard, whose lifelong fascination with rockets began when he was a teenager. He began serious experiments in 1913 after graduating from college.

Q *Which lunar crater is named after a French wine?*

A St. Georges. Geologist Jack Schmitt had bet Dave Scott and Jim Irwin that they would find volcanic crystalline rocks around a particular crater. The wager was several bottles of Nuits St. Georges, the same wine Jules Verne's lunar voyagers had used to toast their success. The crater was later named after the wine.

Q *What was the name of the command service module spacecraft for the Apollo 9 mission?*

A *Gumdrop.*

Q *What is "Go" fever?*

A The mood that sweeps over the technicians, officials, and support personnel at the Cape prior to any launch. It is detected by the feeling of excitement that can be heard in peoples' voices.

Q *What was Big Joe?*

A The first Mercury-Atlas test launch, September 9, 1959.

Q *What was the term "the Gold Team" intended to refer to in connection with Project Mercury?*

A *Life* magazine attempted to get NASA to designate the first three astronauts in space as the Gold Team. The other four were to be called the Red Team. NASA declined to use the designations for fear they might create unnecessary hostilities among the seven. The proposed Gold Team, seen in this May 1961 photo, was: John Glenn, Gus Grissom, Alan Shepard.

Q *What was the Tiger Team?*

A The team of flight controllers and engineers assigned to the tricky re-entry phase of Apollo 13. It consisted of the White Team plus extra technicians and "brass."

Q *What did the "7" signify in the call sign of the Mercury spacecraft?*

A Al Shepard set the "7" tradition by naming his craft *Freedom 7.* Although the media reported that it stood for the seven astronauts, he says it was capsule 7 atop booster 7 for the first of 7 suborbital flights. What he doesn't say is that the pilots are also notoriously superstitious.

Q *Who was Mercury in classical mythology?*

A The son of Jupiter, known for traveling swiftly between Heaven and Earth to intercede with the gods on the behalf of man. In addition to being an inventor, he was patron saint of merchants, travelers, gamblers, thieves, and diplomats, which seems to take in almost everybody in the space program.

Q *What was the Flight of the Classics?*

A The Salyut 6 mission of Georgi Grechko and Yuri Romanenko, in late 1977. Their names are the Ukrainian words for "Greek" and "Roman."

Q *What was Old Number 7?*

A The Russian R-7 rocket, nicknamed *Semyorka,* built to carry a 2-ton nuclear warhead to the U.S. It became the basis of future Soviet space boosters.

Q *Which spacecraft was known unofficially as "Phoenix"?*

A Apollo 7, suggesting that the program was returning to life out of the ashes of Apollo 1.

Q *What was the Phantom Agena?*

A An imaginary target vehicle that Gemini 5 chased through the heavens as a rehearsal for orbital rendezvous.

Q *Who was the Phantom Cosmonaut?*

A Vladimir Illyushin. According to a phony story printed in the London *Daily Worker,* the son of the famed aircraft designer flew 3 orbits on April 7, 1961, making him the first man in space, but was injured in a hard landing and taken to China for hospitalization.

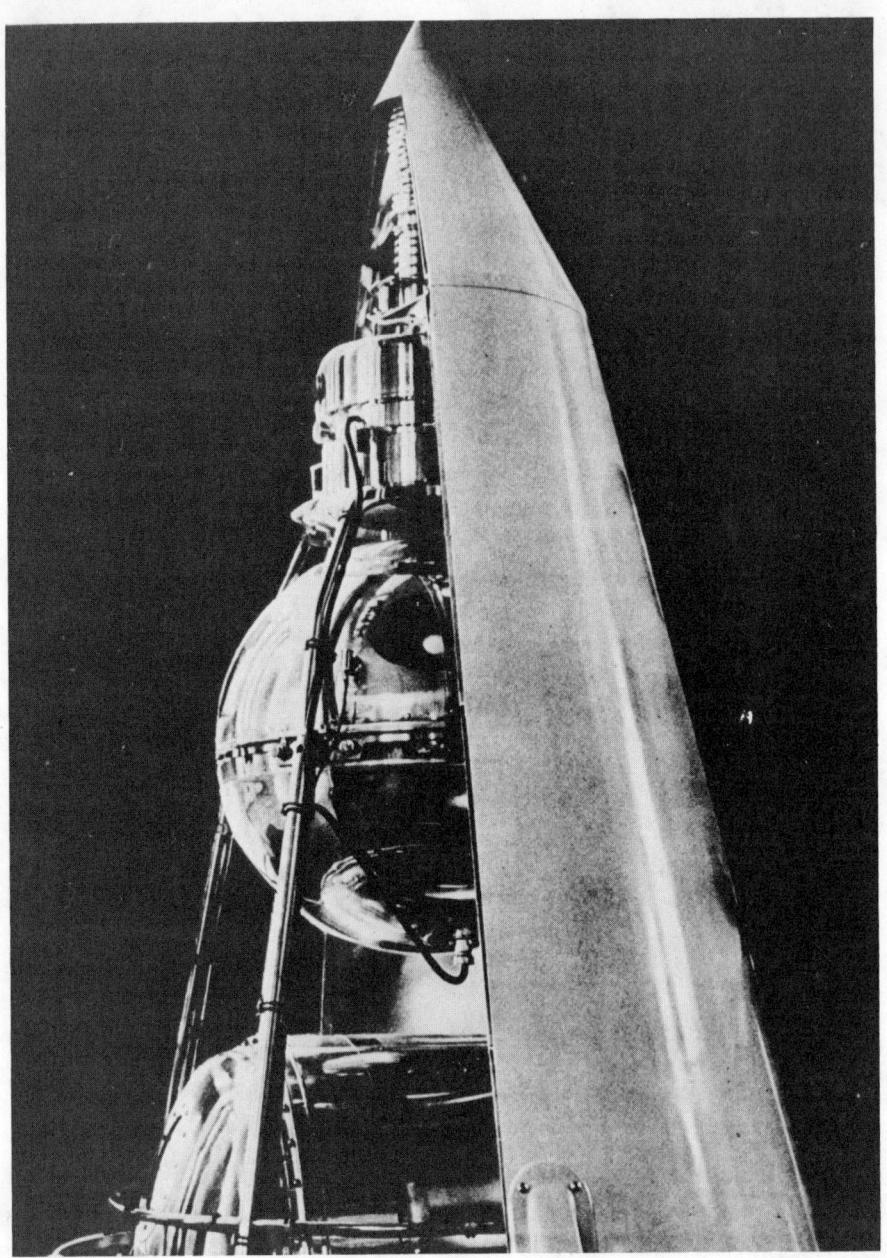

Q *What does* Sputnik *mean?*
A The English translation of the Russian word is, roughly, "traveling companion." This cut-away model shows the Soviets' second satellite as it looked after launch but before orbit insertion.

> **Nom de Plume** Every new year in the early 1960s, *Pravda* would print an essay by Prof. K. Sergeyevich discussing what was planned in the space program for that year. After his death, it was revealed that the annual letter actually came from Sergei P. Korolyov, head of the Soviet space program.

Q *To what project did the names Kaputnik, Stayputnik, and Flopnik refer?*

A British newspapers used the nicknames for the Navy's troubled Vanguard missile.

Q *What did American scientists call the Soviet Sputnik 2, which carried a dog into orbit?*

A Muttnik.

Q *What does "astronaut" mean?*

A "Star voyager." It was chosen in part for its similarity to "aeronaut," by which eighteenth century balloonists were called.

Q *What did the Americans call the first Cuban in space?*

A A Castronaut.

Q *Who was "the voice of Mercury Control"?*

A John A. (Shorty) Powers, the NASA official who not only briefed the press about the technical and scientific aspects of each flight but also was the voice most Americans identify with the early days of the space program. Powers was NASA's link to the world starting with Alan Shepard's flight.

Q *Who is "Stoney"?*

A The astronaut assigned as CapCom at Kennedy Space Center, who talks the crew through launch preparations and lift-off. The origin of the nickname has been forgotten.

Q *Which astronaut was aboard the spacecraft named* Freedom 7?

A Alan Shepard. It was the first Mercury-Redstone manned flight, on May 5, 1961.

Q *What was the name of the rocket sled at White Sands?*

A *Sonic Wind One.*

Q *Why was the name of the first shuttle changed from* Constitution?

A Thousands of *Star Trek* fans swamped Washington with petitions demanding the name be changed to *Enterprise*. It was done with glitter and flourishes, but the Trekkies fell for the old Corbomite Maneuver: after approach and landing tests, it was discovered that refitting *Enterprise* for spaceflight would be too expensive and would make the ship too heavy, so it was cannibalized for spare parts. Some genuine Trekkies were at Palmdale, California, on September 17, 1976, however, when the *Enterprise* officially went on public view. Seen with NASA administrator Dr. James E. Fletcher, from the left: DeForest Kelley (Dr. McCoy), George Takei (Mr. Sulu), Nichelle Nichols (Lt. Uhura), Leonard Nimoy (Mr. Spock), Gene Roddenberry (Star Trek producer), and Walter Koenig (Ens. Chekov).

Q *Which shuttle astronaut was compared to Mr. Spock?*

A Dr. Shannon Lucid's husband, Michael, explained to their three small children her role of mission specialist by saying, "Your mommy might be like Mr. Spock on *Star Trek*." In this survival training photo, however, she looks more like an aquatic football player. Training was at Homestead Air Force Water Survival School in Florida.

Q *Which astronaut was known as "Dr. Rendezvous"?*

A Buzz Aldrin, who did his doctoral thesis at M.I.T. on the factors involved in achieving rendezvous between two orbiting spacecraft. A member of the third group of astronauts selected, Aldrin is seen here having a rendezvous with a robot arm at the Nuclear Rocket Development Station in 1964. The robot arm is used to disassemble nuclear reactors after they have been "hot" tested.

What's in a Name?   The shuttle *Atlantis* is named not for the fabled lost continent but for Dr. Robert Frosch's research ship at Woods Hole Oceanographic Institution. Still, *Atlantis* is considered an unlucky name. With a tricky landing site at the edge of the Bermuda Triangle, would you want to pilot a craft with the same name as something that sank beneath the sea without a trace?

Q *What were the original names for the space shuttles?*

A *Constitution, Columbia, Challenger, Discovery,* and *Atlantis.*

Q *Who selected the names for the first fleet of space shuttles?*

A Administrator Robert Frosch. His original suggestion to use the names of British warships was vetoed by President Carter. The names finally chosen were those of American warships and exploration vessels.

Q *What city was almost renamed "Spacetown"?*

A East Derry, New Hampshire, Alan Shepard's home.

Q *What is Star Town?*

A The Soviet Mission Control and cosmonaut training area, 45 miles northeast of Moscow. Established in 1960, it is often called Star City in the Western media, leading to confusion with another Soviet facility. The cosmonaut training center was renamed for Yuri Gagarin after his death in 1968, and his untouched desk is kept there as a shrine.

Q *When did Cape Canaveral become Cape Kennedy?*

A President Lyndon Johnson ordered the name of NASA's Merritt Island Launch Facility changed to the John F. Kennedy Space Center in 1964. But he also ordered the name of the whole area changed, which didn't sit well with local residents and which caused confusion between the Kennedy Space Center and the Cape Kennedy Air Force Station, site of the Mercury and Gemini launches. The confusion continues to this day. In 1973 the town of Cape Canaveral returned to its old name, and the Air Force station reverted to its older form: Cape Canaveral Air Force Station.

Q *To what do the names Disneyland East and Malfunction Junction refer?*

A Cape Canaveral, during the "missile gap" catch-up days of 1958–61. It was an era of spectacular explosions during launch.

Q *Which of the Apollo flights was dubbed "a connoisseur's mission"?*
A Apollo 9, by astronaut Rusty Schweickart, who with James McDivitt and David Scott was in space from March 3 to 13, 1969. The mission was a complete success in all respects. He is seen here in Panama during survival training.

> **I Know, I Know** Following the Apollo 11 launch, an Italian reporter had a hard time making flight reservations home. In halting English he had to speak to a reservations clerk at Orlando Airport, explaining that his name really was Orlando Orlando.

Q *Which town was known to the astronauts as the City of Lights?*

A The city of Perth, Australia, earned the name when residents began the custom of lighting up the town during flyovers, starting with the John Glenn mission.

Q *Who was known as "der Pad Fuehrer"?*

A Kennedy Space Center pad manager Guenter Wendt, in charge of launch site operations. He was often the last person astronauts saw before being sealed into their spacecraft.

Q *What was the Golden Trough?*

A The executive dining room at North American Rockwell's plant at Downey, California, where astros were given carte blanche; some pigged out on several shrimp cocktails per meal.

Q *Who were the Trench Officers?*

A The FIDO (flight dynamics officer), GUIDO (guidance officer), and RETRO (retrofire officer), who sat in the front row of consoles in Mission Control. They were among the highest ranking in Houston hierarchy and spoke of "the Trench" as if it were a private club.

Q *What significance do the colors white, black, maroon, and gold have for the Apollo lunar program?*

A Those were the names given to the four shifts of officials and engineers assigned to the Mission Operations Control Room.

Q *Where did the sign* SHEPARD AND GRISSOM EXPRESS *appear?*

A It was hung in the air-conditioned van that transported Gus Grissom from his quarters to the pad for the launch of *Liberty Bell 7*.

Q *Identify the astronaut who named his spacecraft* Sigma 7. *What did it mean?*

A Wally Schirra's Project Mercury flight was *Sigma 7*. He selected the name because it was an engineering symbol that stood for the solution to a problem. His flight was considered "a textbook flight."

Q *Which astronaut was called by reporters "John Glenn's little brother"?*
A Fellow Ohioan Neil Armstrong, less for the physical resemblance than for the similarity in backgrounds and WASP identity. In this photo the man who became the first American to orbit the Earth and the man who would be the first human to set foot on the Moon relax during survival training in Panama in June 1963.

> **Son Visits Namesake** Buzz Aldrin, who landed on the lunar surface as part of the Apollo 11 crew in July 1969, went from Moon to Moon. His mother's name was Marian Moon.

Q *Who sent the telegram from Buckingham Palace congratulating the Apollo 11 crew after their splashdown?*

A Col. Armstrong Aldrin Collins of the Queen's Household Guard.

Q *What was the Spirit of Apollo?*

A The flush of pro-American sentiment after the first Moonwalk, and the belief that Americans can solve any problem with enough effort. President Nixon invoked that spirit many times on a worldwide public relations tour, and it was even suggested the presidential airplane be christened "Spirit of Apollo."

Q *What was the Ten-Day Cold Capsule?*

A Apollo 7. Wally Schirra, Walt Cunningham and Donn Eisele came down with colds.

Q *Who was nicknamed "Tweety Bird"?*

A Charles (Pete) Conrad.

Q *Identify the shuttle astronaut nicknamed "Pinky."*

A Dr. George Nelson, the NASA astronaut who attempted the first retrieval of the Solar Max satellite, April 8, 1984. Nelson says he was tagged with the nickname in infancy.

Q *What was the name of Yuri Gagarin's Vostok 1 spacecraft?*

A *Kedr* ("Cedar"). It was unusual because the other five Vostok craft were named for birds. Some sources say the others were named for birds because "Gagarin" is Russian for "wild duck."

Q *Which astronaut was nicknamed "Hoot"?*

A Group 8 astronaut Robert Gibson, after the old Hollywood cowboy.

Q *Who is the only person assigned to talk to astronauts in space?*

A The CapCom, or capsule communicator, who is also an astronaut. Since they've trained together, the CapCom may recognize the significance of each crew members' tone of voice or speech pattern, which a nonastronaut might miss. It also preserves the apocryphal tradition that nobody gives orders to astronauts except another astronaut.

Q *What was a Nixon Fellowship?*

A Being on a Nixon Fellowship was a biting euphemism for being unemployed, after Nixon scrubbed the Apollo program. Few other job descriptions called for the ability to put men and equipment on the Moon, unless one also spoke Russian. Many scientists and technicians spent years relocating. In this 1958 photo, then Vice President Richard M. Nixon visits scientists at the Jet Propulsion Laboratory to congratulate them on the successful launch of the free world's first satellite, Explorer 1.

Q *What was the official name for Alan Shepard's Project Mercury flight in May 1961?*

A It was simply logged in the records at Cape Canaveral as "Test No. 108." In this photo Shepard is honored after the historic flight in a White House ceremony presided over by President John F. Kennedy.

Q *It was nicknamed* Spider *and it was the first test in space of the lunar module. During which Apollo mission did the test take place?*

A Apollo 9, March 3–13, 1969. Jim McDivitt and Rusty Schweickart tested it while David Scott remained in the command module. This photo shows the docked Apollo 9 command service module, *Gumdrop*, and lunar excursion module, *Spider*, in orbit. The Earth is seen clearly in the background.

Pretty as a Picture More than 2,400 photographs of Earth were taken from space during Project Gemini. They were taken from distances ranging from 100 to 850 miles above Earth.

Q *What was the James E. Webb Memorial Rocket?*

A The giant rocket the Soviets were supposedly building to land men on the Moon, and which NASA administrator Webb constantly referred to in speeches in an effort to prompt America to beat the Russians.

Q *Who was Cosmonaut #2?*

A Gherman Titov. In early versions of Gagarin's memoir, *Road to the Cosmos*, Titov's name is not revealed. The Soviets don't like to announce the name of a cosmonaut until the mission is under way.

Q *What are the One Hundred Steps of Glory?*

A The red-carpet reception for cosmonauts after recovery. At the end of the hundred steps, the commander formally reports the success of the mission to the first secretary of the Communist Party or other high Soviet officials.

Q *Name the Cocoa Beach, Florida, hotelman who was known as "the innkeeper to the astronauts."*

A Henri Landiwirth, who operated the Holiday Inn on Route A1A north of Cocoa Beach. This was the motel most frequented by the Original 7.

Q *Which spacecraft were almost named for rock concert halls?*

A Record executive Mickey Kapp suggested that NASA reach the youth of America by naming the Apollo 15 craft "Fillmore East" and "Fillmore West" after the two auditoriums which had recently closed. Kapp had also suggested, with partial success, that Apollo 11 be named "Eagle" and "Butterfly."

Q *What did the letters in MASTIF stand for?*

A Multiple Axis Space Test Inertia Facility. Many of the astronauts agreed that this device gave them their wildest rides in training. It was not used in training after Project Mercury.

Q *What is the Devil's Merry-Go-Round?*

A The centrifuge at the Gagarin Training Center. A rotator, similar to a hamster cage on high speed, has been dubbed the Iron Maiden.

Q *What was known as the "Gusmobile"?*
A Because he had contributed so much to its development at McDonnell-Douglas, technicians and NASA personnel affectionately called the Gemini spacecraft the "Gusmobile" in honor of astronaut Gus Grissom. Grissom, surrounded by some of the other sixteen astronauts during survival training in Panama in 1963, plays Napoleon in the palm rain-hat he and the others were taught to make.

> **Fast Photography**   Gemini 9 photographed slightly more than three fourths of Peru in just under three minutes during its flight in June 1966. Until then, forty years had been spent mapping the country by aerial photography, and not more than one quarter of the task had been completed.

Q *What name did Gus Grissom originally suggest for the Gemini 3 spacecraft?*

A "Wapasha," after the Indian tribe for whom the Wabash River is named. He decided against it when he realized the media would probably call it the "Wabash Cannonball." Instead Gemini was named *Molly Brown* after a Broadway play, to defuse jokes about his first craft sinking. NASA disliked the name and asked for another choice. He said, "What about *Titanic?*" *Molly Brown* stuck.

Q *Name the member of the Original 7 nicknamed "Rah-rah."*

A Wally Schirra. The name comes from his days at Annapolis.

Q *Where is the constellation Urion?*

A Wherever humankind goes into space. It's Wally Schirra's name for the scattering of ice crystals formed after urine is jettisoned.

Q *What are Lunar Pampers?*

A As the name suggests, an "absorbent undergarment" used on long Moonwalks

Q *What is DACT?*

A Disposable Absorbent Collection Trunks, the shuttle era's successor to Lunar Pampers, for use by female astronauts in their pressure suits. Mercifully, the suits aren't often worn. Male liquid elimination is handled by directly attached hoses, officially sized small, medium, and large, but known as extra large, immense, and unbelievable.

Q *What is the call name for Soviet Mission Control?*

A *Zarya* ("Dawn"). The name was first used for the original center in the Crimea, and now for the current center near Kaliningrad.

Q *For what was the Salyut space station named?*

A The first version was launched in April 1971, ten years after Gagarin's historic flight. It was a "salute" to the late Gagarin.

Q *Who was known as "Mumbles"?*

A Soft-spoken Oklahoman Tom Stafford. The name dates from his days as a flight-test instructor at Edwards.

Q *Identify the spacecraft and command module named* Charlie Brown *and* Snoopy.

A Apollo 10's command module was *Charlie Brown* and its LM was *Snoopy.* They flew the LM to within 9.4 miles of the Moon's surface. The crew was Tom Stafford, Gene Cernan, and John Young. Here Stafford examines *Snoopy,* the symbol of NASA's Manned Flight Awareness Program, prior to launch. The Snoopy Awareness Program was designed to promote high-quality workmanship and performance by contractors and NASA personnel.

> **Were They Really Going for It?** According to cosmonaut statements in the 1960s, the Soviets planned a two-vessel, six-man lunar landing mission. The landing team would have included Yuri Gagarin and Valentina Tereshkova, the first man and woman in space, and possibly a "guest cosmonaut" from another Soviet bloc country. It was no secret that the Kremlin wanted a landing to coincide with the fiftieth anniversary of the Bolshevik Revolution, in 1967.

Q *What is a Clarke Orbit?*

A Also called a geosynchronous or geostationary orbit, it gives the spacecraft the appearance of hovering over the same spot. Back in 1945, British signal corpsman Arthur C. Clarke calculated the altitude at which a satellite's orbital speed would match Earth's rotational speed. The 22,300-mile altitude is now widely used by communications satellites. Were it possible to patent ideas, Clarke would be a rich man today. He wrote a story about it, titled "How I Lost a Billion Dollars in My Spare Time."

Q *What was a "chaperone"?*

A During the Gemini and Apollo programs, it was a specific engineer assigned by the contractor for each spacecraft part. The chaperone was responsible for production, checkout, transport, installation, and final testing. Along with the responsibility went the power to buck foremen or the board of directors—even to stop the assembly lines.

Q *Give the call signs of either of the two Voskhod spacecraft.*

A The three-man Voskhod 1 was called *Rubin* ("Ruby"), and the two-man Voskhod 2 was called *Almaz* ("Diamond").

Q *What did the names of the Soviet manned programs mean?*

A *Vostok* means "East," *Voskhod* means "Ascent," and *Salyut* means "Salute," as in an artillery salute. *Soyuz* means "Union," and was the first dockable Soviet spacecraft.

Q *What was the Goose Rule?*

A The rule that during goose migration season, there must be two pilots in the T-38 trainer. The rule made it impossible for the nonpilot astronauts in Group 6 to get their required flight time, and the term became a NASA synonym for Catch-22. The rule was enacted after Theodore Freeman was killed in a plane crash after a goose slammed into his cockpit on October 31, 1964.

Q *Who was Lt. Col. William Douglas?*

A The physician for the Original 7 Mercury Astronauts, seen here having breakfast with them on the morning of Alan Shepard's suborbital flight in 1961. Douglas is at the left, at the head of the table.

Q *Who was "007"?*

A The security chief of Kennedy Space Center, Charles Buckley, whose sensitive position and dealings with international celebrities led his friends to get him that Federal Employee Identification Number, the only "00" number ever given by the U.S. government. Buckley (third from right) joined John Young (right) and Deke Slayton (between them) and other officials at a prelaunch breakfast for Apollo 16 in April 1972.

Q *What is the Crawler?*

A The special tractor that had to be invented for the purpose of moving Apollo-Saturn spacecraft from the Vehicle Assembly Building at Kennedy Space Center to the launchpad. It requires a crew of thirteen, including three drivers, and travels at 1 mph when loaded. Each of the shoes on the eight treads weighs 1 ton.

> **Red Moon for Christmas** The Soviets may have tried to launch a manned Moon mission just before Apollo 8 was launched. There's evidence a one-man fly-around was planned for December 9, 1968, but was scrubbed at the last minute for reasons yet unknown. Pavel Belyayev, the cosmonaut believed assigned to that mission, died in surgery the following year and was the first cosmonaut not to be buried in the Kremlin wall.

Q *What piece of space hardware is known as the "Polish Bomber"?*

A The space shuttle, because the "bomb-bay doors" open upward. Among other nicknames for the world's largest spacecraft: the Great Brick Bird, the Flying Brickyard, the Great White Batmobile, the Space Dildo, the Intergalactic Hilton, and Capricorn One.

Q *By what name was the LANDSAT spacecraft originally known?*

A Earth Resources Observation Satellite (ERTS). The change was made when the government decided it didn't want to have to explain "Project EROS."

Q *What is BGT?*

A Basic Grubby Training, the 240 hours of classroom instruction that the Group 3 astronauts received after their selection. Al Shepard even put together a BGT diploma for each man, granting him permission to leave Houston without a den mother.

Q *What was the Pregnant Guppy?*

A A Boeing Stratocruiser modified to ferry segments of the Saturn booster during the 1960s. An even larger version was christened Very Pregnant Guppy, but that was soon replaced by the more macho label Super Guppy.

Q *Which pro sports team was named after American space travelers?*

A The Houston Astros. The city also honored its most famous residents by naming its new stadium the Astrodome.

Q *How did the Muroc dry lake, and later the Muroc Air Force Base, come by that name?*

A Although many, including writers who should know better, think it's an Indian word, in fact the area is named for the first Anglo homesteaders, the Corum family. At least one Corum was still on the land when supersonic flight began.

Q *Who nicknamed Project Mercury "Spam in a can"?*

A The test pilots at Edwards Air Force Base, who felt being an astronaut simply meant being a passenger in a space capsule, not the pilot of a spacecraft.

Q *What was Cluster's Last Stand?*

A The launch of SA-1, the first Saturn launch, in October 1961. The vehicle resembled several V-2 type rockets clustered together.

Q *What were nicknamed "The Flying Bedstead" and "The Great Train Wreck"?*

A "The Flying Bedstead" was the mock-up of a lunar module (LM) that the Apollo astronauts trained in. The name referred to its structure and unusual shape. "The Great Train Wreck" was the name given by John Young to the huge, bulky Apollo mission trainer at Kennedy Space Center.

Q *Identify the U.S. aircraft carrier that recovered Gus Grissom and John Young after the first manned Gemini flight.*

A The USS *Intrepid* (CV-11), which had also recovered Scott Carpenter after his *Aurora 7* flight in the Mercury program. She is seen here exiting Norfolk, Virginia, en route to her station for Carpenter's flight.

Q *What is the Panic Button?*

A The remote-control destruct device installed on rockets after the spectacular explosions of the late 1950s. Even the shuttle has a panic button, designed to pulverize the SRBs (solid rocket boosters) before they can strike personnel or buildings should the launch go off course. On most launches, concurrence among three officers is needed for destruct; it usually comes within a second of recognition.

Q *What is lift-off?*

A Officially it means 2 inches of altitude. If an engine shuts down after lift-off, the impact from falling just those few inches can crush the ship. The call of "Lift-off" is largely for the benefit of the astronauts aboard, who can't really tell otherwise.

Q *Which astronauts shared the nickname "José"?*

A Both Al Shepard and John Young have been tagged with the name, a reference to their impersonations of José Jimenez.

Q *What was the Coast Guard Missile?*

A Cape Canaveral's tall, thin, white lighthouse. Rookie birdwatchers (and some reporters) often wasted time and film waiting for it to rise out over the Atlantic.

Q *What is Stable 2 position?*

A The opposite of Stable 1, which means floating right side up. The Apollo command module did poorly in the waves after splashdown, and on Apollo 7 it capsized. "Stable 2," meaning capsized, became a euphemism for "ass-backwards."

Q *What is MOCR?*

A Pronounced MOE-kur the Mission Operations Control Room is the sea of screens, phones, and consoles most TV viewers think of as "Mission Control," a term which covers a much larger area at Johnson Space Center. (There are actually two MOCRs, to handle simultaneous missions.)

Q *What is g-force and who coined the term?*

A A force equal to the pull of normal Earth gravity. The unit is used to measure the effects of acceleration or deceleration on objects or organisms. The term, along with "biodynamics," the study of organisms in motion, was popularized by Dr. John Paul Stapp, the inventor of the rocket sled.

Q *What was the code name of the helicopter that tried unsuccessfully to lift Gus Grissom's Mercury spacecraft out of the water after splashdown?*

A *Hunt Club 1* (Marines #32, *not* seen here). Immediately after the hatch of the spacecraft blew, it began to fill with water, and the added weight nearly pulled the chopper into the water as the craft sank. The copter pilot was Lt. James Lewis. The pilot who actually lifted Grissom was Lt. John Teinhard (Marines #30), and the co-pilot was Lt. George Cox, who also participated in Alan Shepard's recovery. Grissom's spacecraft was the only one the U.S. did not recover in the manned Mercury, Gemini, and Apollo programs.

Q *What was Old 66?*

A The helicopter that retrieved the crews of Apollo missions 8, 10, and 11.

Q *Who termed the X-15 rocket plane "America's first spaceship"?*

A Author Richard Tregaskis, who wrote a book about the X-15 project with that phrase in the title. He was also the author of *Guadalcanal Diary*.

Q *What was the name Chuck Yeager gave to his P-51 in Europe during World War II?*

A *Glamourous Glennis,* after his wife; it was written on the nose of the plane. He also put the same name on the X-1, the plane in which he broke the sound barrier.

Q *Who or what is Just Plain Lucky?*

A NASA's Jet Propulsion Laboratories in Pasadena, so named by workers there because of serendipitous discoveries.

Q *What do NASA insiders mean when they speak of the "other company"?*

A This is the term they use for the Soviet space program, borrowed from the CIA's term for the KGB.

Q *What is the origin of the word "rocket"?*

A The 14th-century Italian historian Muratori used the word *rocchetta,* referring to an implement used in spinning thread, to describe the shape of the weapons introduced by the Arabs.

Q *What did each service branch call its own plan to put a man in space?*

A The Army plan was called Project Adam and was sold as a suborbital troop transport (honest!). The Navy venture was called Manned Earth Reconnaissance (MER), and the Air Force contributed Project MISS (Man in Space Soonest). Elements of Miss and Adam were used in Project Mercury; MER bore a slight resemblance to the modern space shuttle.

Q *What was Card File 23?*

A The code name for the radio relay aircraft that functioned during the recovery of Gus Grissom after the flight of *Liberty Bell 7* in the Project Mercury program.

Q *What did the Group 8 astronauts call themselves?*
A ''TFNG,'' for ''Thirty-Five New Guys.'' Here, Fred Hauck models the TFNG designer collection T-shirt.

Q *What did the Original 7 Astronauts call the private meetings they held among themselves to work out problems?*

A Seances.

Q *What was the subject of discussion at the Kona Kai Seance?*

A The now-famous confrontation at San Diego's Konakai Hotel between John Glenn and other members of the Original 7 was about womanizing. Glenn told the group that such activity could bring unfavorable publicity to Project Mercury. He advised them to keep their zippers shut.

Q *What was the name of the 10-inch-long rod with a hook on the end that the early astronauts used to reach certain levers and buttons in their capsule during flight?*

A The Swizzle Stick. Many astronauts regarded it as an extension of their own fingers.

Q *Which astronaut is known as "Ox," and why?*

A James van Hoften. At 6 foot 4 inches and over 210 pounds, he is the largest astronaut to date.

Q *What does the acronym ALFA stand for, and what is it?*

A ALFA is the Air Lubricated Free Axis Trainer used in early astronaut training. The original name, Air Bearing Orbital Attitude Simulator, was considered a tongue twister, so it was changed.

Q *Who, or what, was SARAH?*

A The Search and Rescue and Homing beacon that directed search planes to a space capsule in the water, awaiting recovery.

Q *What was the Voas-meter?*

A A complicated device for making scientific and navigational measurements outside the Mercury spacecraft. It was named by the astronauts after its inventor, Dr. Robert Voas, who had named it the extinction-spectro-photo-poleriscope-occulo-gravo-gyro-kinetometer.

Q *What was the GIRD?*

A Group for the Investigation of Reaction Dynamics, the Russian think tank that built that nation's first liquid-fuel rocket. Since they received little funding and very low priority, they used the same Russian initials to claim they were the Group Working For Nothing.

Q *What were Buzz Aldrin's first words on setting foot on the lunar surface?*
A "Beautiful! . . . magnificent desolation!"

Messages and Quotes

Q *Who said: "Roger, lift-off, and the clock is started!"?*

A Alan Shepard. It was the first utterance by an American on a journey into space.

Q *What were the first words uttered by Neil Armstrong after leaving the interior of the lunar excursion module on the surface of the Moon?*

A "Okay, Houston, I'm on the porch." This was said outside the spacecraft and on the Moon but before he had gone down the ladder and actually set foot on the surface.

Q *Who said "I have suited up just in case you two chicken out and turn the mission over to the backup team."*

A Wally Schirra, arriving at a last-minute Gemini 3 briefing dressed in the remnants of a Mercury spacesuit.

Q *Who called the space program "Technological life insurance for this country"?*

A Frank Borman, testifying before Congress.

Q *Who said: "The Moon isn't made of green cheese at all. It's American cheese!"*

A William Anders, during the flight of Apollo 8.

Q *Who said: "Columbus was right, the world is round"?*

A John Young, from 467 miles above Earth during the Gemini 10 flight, July 18–21, 1966. Michael Collins was the other crew member. This was the highest penetration into space to date by humans.

Q *Who said: "We both landed at the same time"?*

A Buzz Aldrin. It is his standard retort to being introduced as the "second American to land on the moon."

Q *Exactly what did Neil Armstrong say when his foot first touched the surface of the Moon?*

A "That's one small step for a man, one giant leap for mankind." Static in the transmission to Earth led many people to believe they heard the historic message without the article "a" before the word "man," and it was reported in much of the media as "step for man." It was 10:56 Eastern Daylight Time (109 hours, 24 minutes, and 20 seconds in Ground Elapsed Time of the mission) on July 20, 1969. Here Apollo 11 crew members Collins, Armstrong, and Aldrin meet with U.S. Postmaster General Winton M. Blount to inspect an enlargement of the airmail commemorative stamp unveiled after their Moon mission.

> **Q** Who said: "The Apollo 15 fellows were crucified. . . . NASA was headline-grabbing so as not to spoil the Boy Scout image"?
>
> **A** Astronaut Ed Mitchell, referring to the controversy that resulted when the Apollo 15 crew carried a batch of autographed stamped cover envelopes to the Moon for later resale on Earth. NASA confiscated the covers after the crew gave one hundred to a German stamp dealer who sold them for $1,500 apiece. Each envelope is postmarked twice: launch day, and aboard the aircraft carrier USS *Ranger* after recovery. Though the envelopes say the series was limited to 300, the former astronauts won an eleven-year court battle for possession of the mementoes in 1983 and NASA returned 359 envelopes to them. Stamp dealers value the lot at approximately $500,000. The photo shown was originally released by NASA in 1972.

Q *"Please be informed there is a Santa Claus."* Who said it?

A Jim Lovell, after he and fellow Apollo 8 crew members Frank Borman and Bill Anders fired rockets to start the service propulsion system needed to send them heading back toward Earth. They had just made their final pass behind the Moon on Christmas Eve, 1968, and if the rockets had failed to fire they would have remained in lunar orbit forever.

Q *"Godspeed, John Glenn."* Who said it? When?

A Fellow astronaut Scott Carpenter, at the moment of engine sequence for the lift-off of *Friendship 7*. Carpenter was the CapCom for the flight. Glenn did not hear the message at the time.

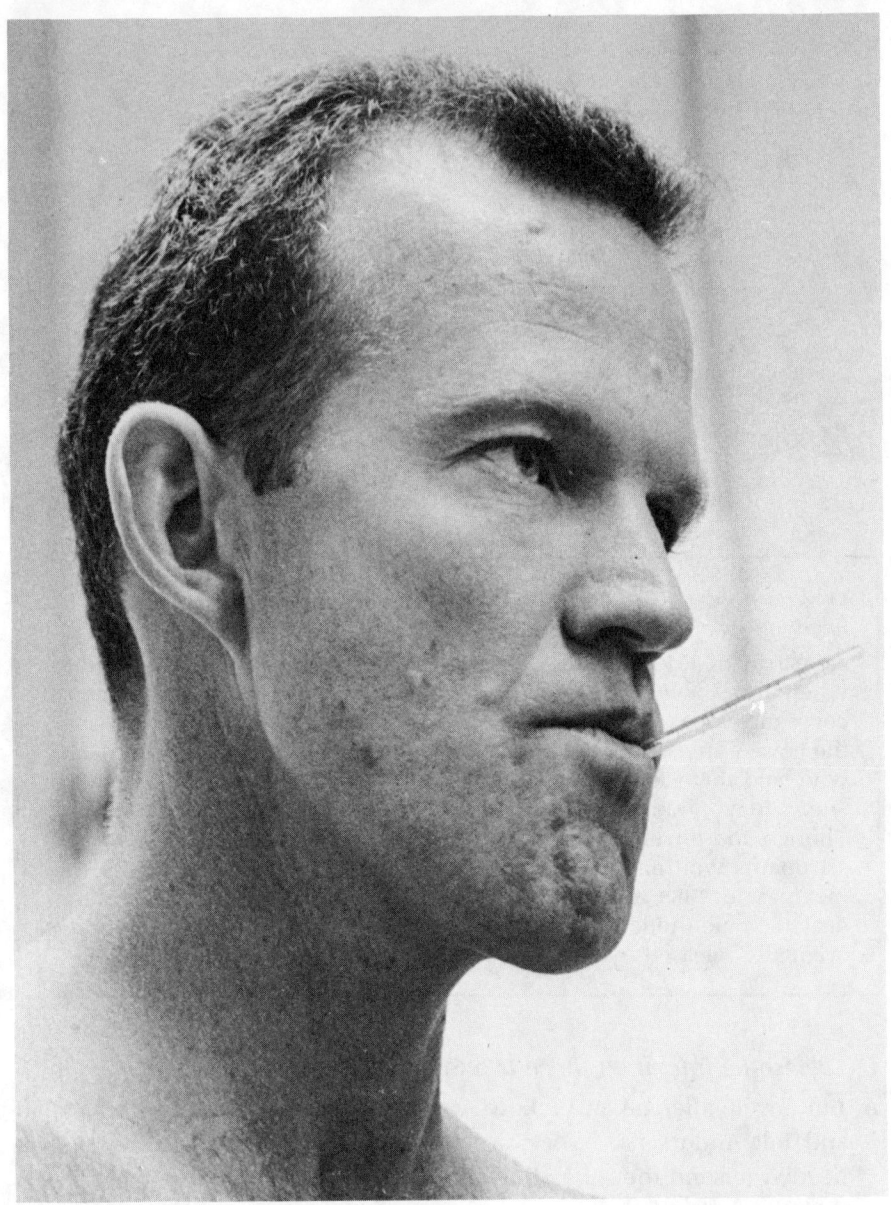

Q *"On the old gazoo." Who said it? What did it mean?*

A Gordon Cooper used the expression to indicate that he had lined up his *Faith 7* Mercury spacecraft for re-entry and recovery. His accuracy allowed him to land less than 4 miles from the recovery ship USS *Kearsarge*. Cooper is seen here undergoing a medical checkup prior to his scheduled 22-orbit Mercury flight.

Q *What sign awaited the Gemini 12 crew as they arrived at the pad?*

A LAST CHANCE! NO RERUNS! SHOW WILL CLOSE AFTER THIS PERFORMANCE! It was the final Gemini mission. Across the seats of their flight suits, Jim Lovell and Buzz Aldrin carried another message as they walked to their spacecraft: THE and END.

Q *"We're tumbling end over end." Who said it?*

A The crew of Gemini 8, Neil Armstrong and Dave Scott. Approximately half an hour after they linked up in space with an Agena rocket, their capsule and the rocket began tumbling wildly. Armstrong was forced to back the Gemini away from the Agena, but then the Gemini began tumbling faster (a complete turn per second) until he fired the rockets intended to line up the spacecraft before re-entry. This ended the tumbling but caused termination of the mission.

Q *Who said: "You look down there and you get homesick. You want some sunshine, fresh air, you want to wander in the woods."*

A Cosmonaut Vladislav Volkov, just before fatal re-entry.

Q *Who said: "I was thinking that everything in here was supplied by the lowest bidder"?*

A Alan Shepard, asked what went through his mind during his 1961 flight.

Q *Who said: "If you can't be good, be colorful"?*
A Pete Conrad.

Q *Who said: "Houston . . . I've got the world in my window"?*
A Mike Collins.

Q *Who said: "I was a rotten SOB before I left . . . now I'm just an SOB"?*

A Alan Shepard, after becoming the first American to fly in space.

Q *Who said, "The idea of becoming the world's oldest, used, permanent-training astronaut does not strike me as very good career planning"?*

A John Glenn, upon announcing his retirement from the Astronaut Corps.

Q *Who said: "I would give my left arm to be the first man in space"?*
A Deke Slayton.

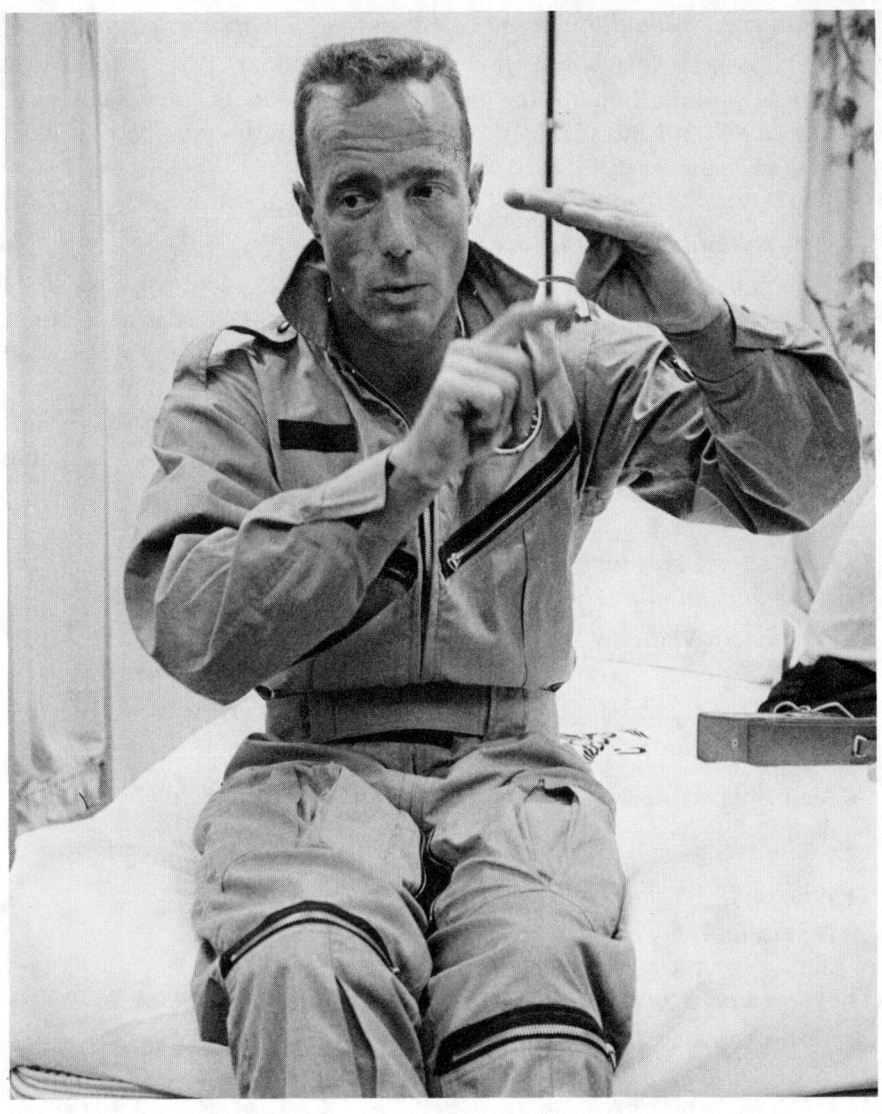

Q *Which reporter said: "I'm afraid we may have lost an astronaut" after a troublesome Mercury flight?*

A Walter Cronkite, after NASA lost contact with Scott Carpenter for almost four hours after his 1962 re-entry. There had been concern that Carpenter, who was critically low on fuel, may have entered the atmosphere too steeply and burned up. In fact, he was in good health but about 250 miles off target. Here, an animated Carpenter describes part of his mission during de-briefing and medical examinations on Grand Turk Island.

> **Don't Try It!** In the 1952 British movie *Breaking the Sound Barrier*, the feat was accomplished by the pilot's reversing the controls at transonic speed. Chuck Yeager, the test pilot who actually did go faster than sound, was asked by the secretary of the Air Force if that is what he did on October 14, 1947. "No, sir, that is not correct. Anyone who reversed the controls going transonic would be dead."

Q *Who said: "We're all asleep down here"?*

A Col. Shorty Powers at Langley, when roused from bed by a reporter asking for a comment on the historic first manned spaceflight by Soviet cosmonaut Yuri Gagarin.

Q *"Now let the other countries try to catch us." Who said it?*

A Soviet cosmonaut Yuri Gagarin, to Premier Nikita Khrushchev while reviewing a parade honoring the cosmonaut in Red Square shortly after his flight.

Q *Who said: "There won't be any flying to do—a monkey's going to make the first flight"?*

A Famed test pilot Chuck Yeager, explaining why he wasn't disappointed about not going into the Project Mercury program.

Q *Who said: "I, for one, don't want to go to bed by the light of a Communist moon"?*

A Vice President Lyndon B. Johnson, on May 11, 1963, in response to complaints by some members of Congress that space race costs were getting out of hand with the new Gemini and Apollo programs. LBJ's support for the space program is considered one of the reasons it received necessary funding in the 1960s.

Q *"The Earth is the cradle of the mind, but one cannot live in a cradle forever." Who said it?*

A Konstantin Tsiolkovsky, "the father of spaceflight," who published articles in Russia on rocketry and traveling in space from 1903 until his death in 1935. He proclaimed that mankind would "set foot on the soil of the asteroids."

Q *Who referred to the flight of Apollo 9 as "ten days that thrilled the world"?*

A President Richard M. Nixon.

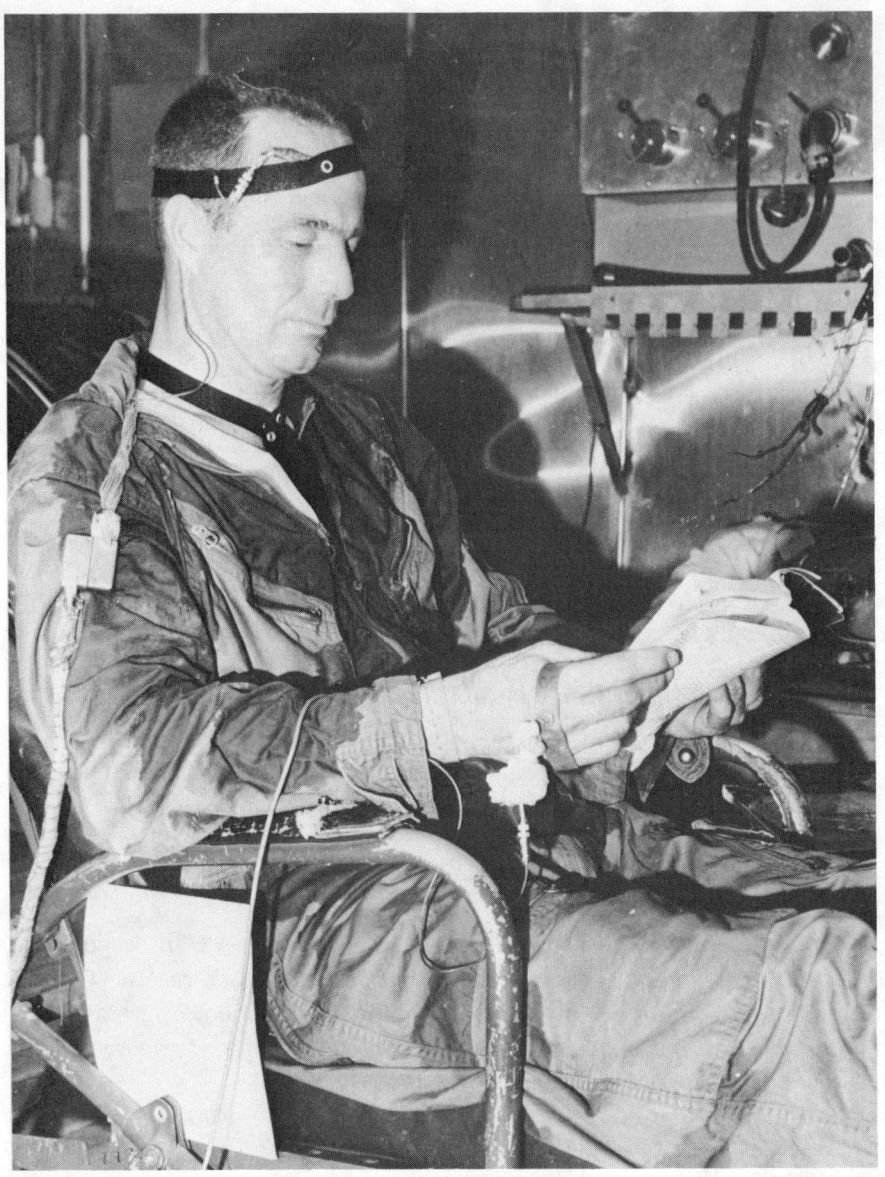

Q *Name the astronaut who sent the first message in Spanish back to Earth.*

A Scott Carpenter, who directed a thirty-seven-word greeting to Mexico as he passed over that country on the first orbit of *Aurora 7*. Carpenter is seen here spending his time reading while participating in a heat-stress test during training. The temperature was 130° F in this chamber at the Wright Development Center, Ohio.

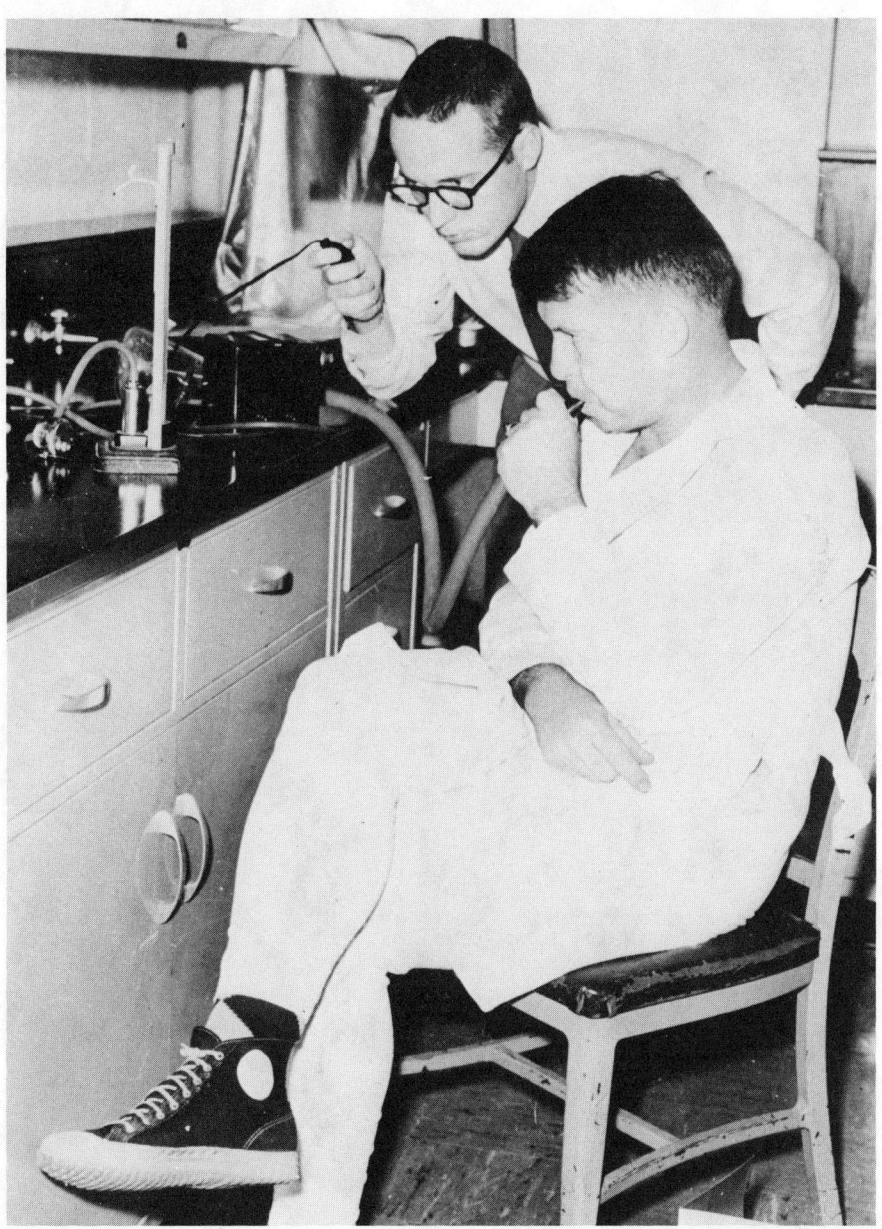

Q *While flying over South America an astronaut radioed "Buenos dias, you all" back to Earth. Who was it?*

A Wally Schirra, on *Sigma 7* in Project Mercury. He was asked to say something in Spanish, which would be broadcast live. Schirra is seen here undergoing a lung capacity test in preflight training.

Q *Who said: "I think Isaac Newton is doing most of the driving now"?*
A Bill Anders, at the left, during Apollo 8, when asked by Michael
Collins, capsule communicator, who was "doing the driving up there."
Collins relayed the question from Collins's son. Anders and fellow
crewmen, James Lovell, center, and Frank Borman, paused for this
photo during training for their 1968 flight.

Big Arms? During the launch of Gemini 2, the rocket shut down seconds after ignition and the booster settled back onto the pad. Bill Hampton of the Cape Kennedy Space Network, convinced he couldn't ad lib coverage, had written out a description of lift-off and was still reading from it when an associate pointed to the Titan still sitting on the pad. His recovery: "Ladies and gentlemen, you won't believe it, but a pair of giant hands have just come out of the sky and pushed that rocket back down onto the pad." Pretty good for a guy who couldn't ad lib.

Q *What was known as "the day we launched the escape tower"?*

A It is a reference to the abortive first test of a Mercury capsule and a Redstone rocket, which resulted in only the escape tower lifting off. The date was November 21, 1960.

Q *"A-OK." Who said it? What did it mean?*

A Often attributed to Alan Shepard during his Mercury flight, it was actually Shorty Powers, the voice of Mercury Control, who used the old telegraphers' expression over the airwaves. The sharp sound of the letter *A* is believed to overcome static better than the softer *O*.

Q *Who said: "Why don't you fix your little problem . . . and light this candle"?*

A Alan Shepard, in frustration over a hold in the countdown for his Project Mercury spaceflight.

Q *"That was a real fireball, boy." Who said it? When?*

A John Glenn, after observing the burning retro-pack chunks go by his spacecraft window. The first five words appeared as headlines in newspapers around the world after his recovery.

Q *"Do good work!" Who said it? When?*

A Gus Grissom, to an assembled group of technicians at the Convair manufacturing plant in San Diego, California. Convair was the prime contractor for the Atlas rocket and Grissom was on an inspection tour. He was asked to say a few words to the people working on the project. And a few words he said.

Q *Who said: "I am an ordinary Soviet man"?*

A Yuri Gagarin, the first man in space, in denying statements by Russian-Americans that he was related to Prince Mikhail Gagarin.

Q *According to Gus Grissom, which was the most critical part of the mission?*

A "The part between lift-off and splashdown." In this photo a technician completes various systems checks of Grissom's spacecraft some weeks prior to launch.

> **On the Tip of His Tongue, But . . .** In show business, "Break a leg" is an expression meaning "do a good performance." Test pilots sometimes say "Go blow up" to a fellow pilot about to make a flight in a new aircraft. Gus Grissom intended to say it to Alan Shepard on the morning of the first U.S. manned space flight. However, as the tension built for that historic flight, Grissom found himself unable to utter the words.

Q *"Weh-ah! I'm a wetback now." Who said it? Why?*

A Alan Shepard, while waiting through the long holds prior to his launch in *Freedom 7*. Shepard was unable to control his bladder and asked Mission Control if he could urinate in his spacesuit since no provisions had been made for this natural function on what was considered a brief flight. Permission was granted, after NASA took into consideration the effects the urine would have on his body sensors. Because of the reclining position of the astronaut in the Mercury spacecraft, the urine collected near the arc of his back and he responded accordingly.

Q *Who said: "You smeared up my windshield, you dirty dog!"?*

A Jim McDivitt, to fellow Gemini 4 crew member Ed White who, while taking his historic spacewalk, took the opportunity to look into the spacecraft. White apparently left some sort of imprint on the glass, which resulted in McDivitt's good-natured ribbing.

Q *Which Mercury flight sent the following message: "CapCom, this is Astro. Am on the window and the view is great. I can see all the colors and can make out coastlines"?*

A Mercury-Atlas-5. The "astronaut" was the chimp Enos, and the message sent to tracking stations from the spacecraft had been prerecorded and placed onboard to simulate voice contact with a human. The humor of the event did not escape President Kennedy, who reported at a press conference that "the chimpanzee took off at 10:08. He reported that everything is perfect and working well!"

Q *Who said: "Some leaf, some rake" in response to critics who described the space program as "leaf-raking"?*

A Then Vice President Lyndon B. Johnson, who was paraphrasing Winston Churchill's famous "Some chicken. Some neck." Churchill's quote came in response to Adolf Hitler's prediction that Germany would wring Britain's neck like that of a chicken. Johnson made his retort after the successful conclusion of Project Mercury.

Q *Who said: "Flying just isn't my cup of tea"?*

A Brian O'Leary, explaining why he resigned. An astronomer, his was to be the first nonpilot astronaut group. A few days into training, they were told they would have to learn to fly jets after all. O'Leary quit after making his second solo. His offhand remark was picked up by the press. O'Leary (left) and fellow scientist-astronaut candidate Joseph Allen are seen here practicing launch escape system procedures shortly after being selected.

Speaking of Plastics Group 6 astronaut Brian O'Leary was the college roommate of Charles Webb, author of the 1960s best-seller *The Graduate*.

Q *During spaceflight who described himself as "soaring like a bird . . . over a huge colored map"?*

A Colonel Alexei-Leonov, during his spacewalk on the Voskhod II flight, in March 1965.

Q *"She's riding like a dream." Who said it?*

A Wally Schirra, describing the flight of Apollo 7, the first manned mission in the program, October 11–22, 1968.

Q Definition of a Sports Car: A Hedge Against the Male Menopause. *Where did this sign appear? Why?*

A On the blackboard in the astronaut office at Cape Canaveral. It was written by John Glenn in response to being teased for continuing to drive his old beat-up car instead of getting a sports car such as the others had.

Q *Who said: "It looks like an angry alligator out there"?*

A Tom Stafford, during the Gemini 9 mission, when he and Gene Cernan attempted to dock with a target vehicle that had been shot into orbit. The protective shroud had not jettisoned, and as a result its two halves protruded and resembled large jaws.

Q *"Don't get too much sun on GTI." Who wrote it? What did it mean?*

A As John Glenn unrolled the flight plan for his mission aboard *Friendship 7* he discovered that Nancy Lowe, the astronauts' secretary, had included the pleasant message in the tiny scroll. GTI was Grand Turk Island, the location where Glenn would be debriefed after recovery.

Q *"Nothing was wrong, but nothing was right either." Who said this in reference to a postponement of John Glenn's* Friendship 7 *flight?*

A Walt Williams, associate director for spacecraft operations. He was later named deputy director for the Manned Spacecraft Center, and still later deputy director for Mission Requirements and Flight Operations.

Q *Who said: "Levity is appropriate in a dangerous trade."*

A Wally Schirra—more than once.

Q *Who said: "Man is the deciding element. . . . As long as man is able to alter the decision of the machine, we can perform under any known conditions"?*

A Project Mercury flight director Chris Kraft, in response to a question asking why it was necessary to put a man inside a spacecraft since it was believed Ground Control could execute any experiments or movements required.

Q *"Ride 'em, cowboy!" Who said it? When?*

A Pete Conrad, to Richard Gordon during Gemini 11, September 12–15, 1966, as Gordon tied a tether cord to an Agena rocket they had docked with. Gordon's efforts reminded Conrad of a scene from the Old West as he attempted to "rope" the Agena while traveling at 17,500 mph.

Q *Who said: "The glory of being first is not everything"?*

A The second American in space, Gus Grissom, fourth from the left in this photo with fellow members of the Original 7 as they prepare for some jet training flights.

> **Murphy's Law** Air Force captain Edward Murphy was a development engineer assigned to Northrop's Project MX981 at Edwards Air Force Base in 1949. A technician had miswired part of a transducer, causing it to malfunction. An exasperated Capt. Murphy said of the tech, "If there's any way to do it wrong, he will." Northrop's project manager, George Nichols, overheard the remark and tagged it "Murphy's Law." Within a few weeks it was being widely quoted at Edwards in the more familiar version, "If anything can go wrong, it will."

Q *Who said: "The astronaut has been added to the system as a redundant component" when talking about Project Mercury?*

A Various engineers and scientists attending an Armed Forces–National Research Council conference at Woods Hole, Massachusetts, in mid-1960. They were expressing their thoughts about the function of a human in what they considered a totally automated spacecraft system.

Q *Who said: "You guys are doing a commendable job of maintaining radio silence. Since the French stopped shooting at you . . ."?*

A Deke Slayton made the quip to John Young and Michael Collins during Gemini 10. The two spacemen were exceptionally quiet and business-like, and Ground wanted to hear from them more. During the mission the French government conducted a nuclear test in the Pacific, dropping the bomb from a plane. The mushroom cloud and atomic flash occurred while Gemini 10 was on the other side of the Earth.

Q *Who said: "I think of Project Mercury and the open manner in which we are conducting it for the benefit of all as a light in the sky. Aurora also means dawn, in this case the dawn of a new age"?*

A Scott Carpenter, in explaining why he called his spacecraft *Aurora 7*.

Q *Who said: "If you're afraid of wolves, don't go into the woods"?*

A Soviet Cosmonaut Valentina Tereshkova, answering a reporter's question on whether she was afraid during her spaceflight.

Q *"Eight days or bust." Which Gemini crew made this their slogan?*

A Gordon Cooper and Pete Conrad for their Gemini 5 flight, August 21–29, 1965. The significance of the duration was that it corresponded to the length of time necessary for a flight to the Moon.

Q *Who said: "I felt God's presence with me on the Moon."*

A Jim Irwin, explaining why he went into the ministry after resigning.

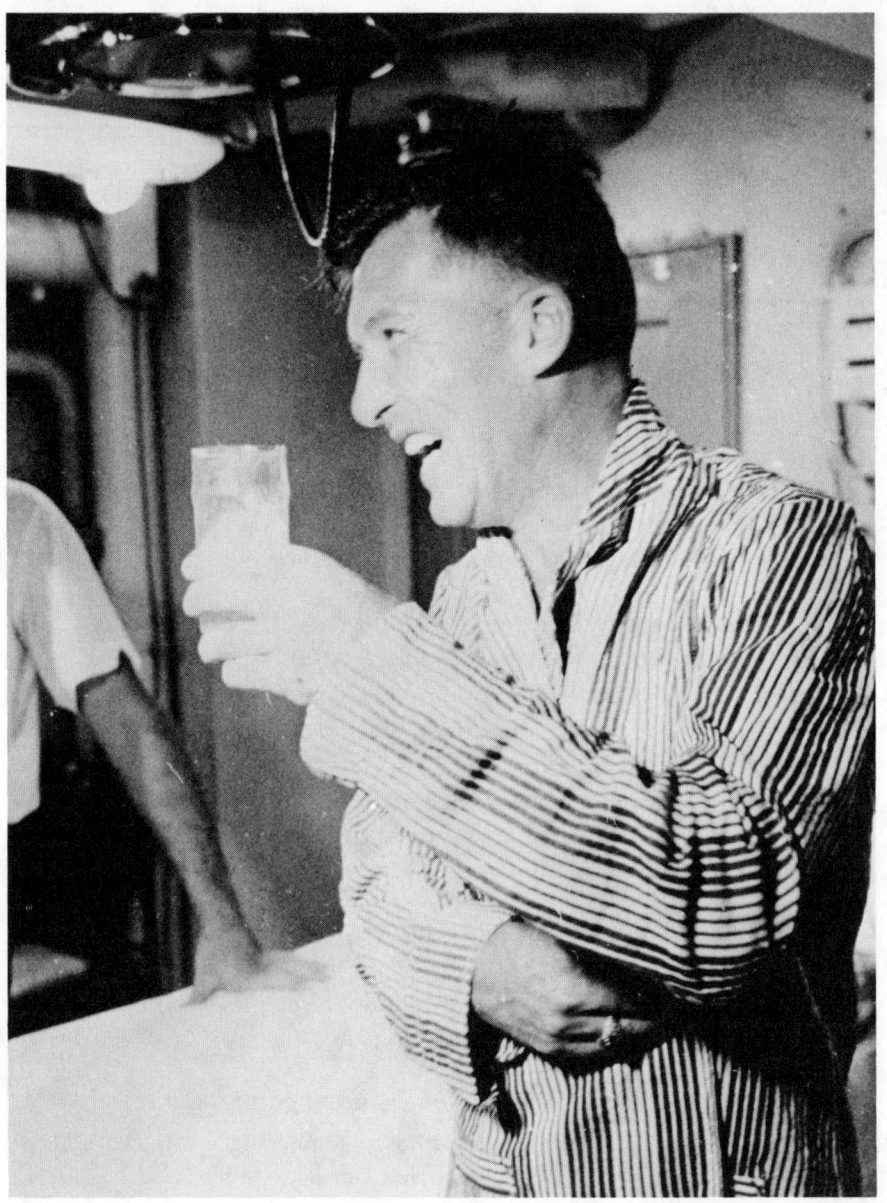

Q *Who said: "This tower is a real sayonara"?*
A Wally Schirra, as he watched the escape tower blow away after launch of his Project Mercury flight, *Sigma 7*, on October 3, 1962. Here an obviously delighted Schirra pauses during his 2½-hour postflight physical to enjoy a refreshing drink.

> **Wrong-Way ABC** To house its news operation for the first shuttle launch, ABC erected a prefab building, which resembled a box set on stilts, at the Launch Complex 39 press site. Jules Bergman, visiting the site a few days before launch, noted with horror that the windows were facing the wrong way. A crane was hurriedly brought in, and the building was picked up and spun around 180°. But just to rub it in, the CBS building next door put up a large arrow, inscribed with the word·"LAUNCH," pointing toward the pad.

Q *"The moon is essentially gray, no color ... looks like plaster of Paris, sort of gray sand." Who said it?*

A Jim Lovell, giving the first close-up description of the lunar surface, during the flight of Apollo 8.

Q *"The President is go for landing." Who said it?*

A An unknown technician in Mission Control Center, who spoke these words on the internal communication loop during the descent of Apollo 11. President Richard M. Nixon was standing by to talk to the astronauts.

Q *"Son of a bitch! What the hell happened?" Who said it?*

A Astronaut Gene Cernan, during the simulated lunar landing of Apollo 10. On LM separation from descent stage, wrong switching caused brief, wild pitching. The words were among the strongest to be heard on live radio at that time, and they marked one of the first times an astronaut had cursed in public.

Q *"We believe that long-lived space stations with interchangeable crews will be humankind's main road to the universe." Who said it?*

A Soviet Premier Leonid Brezhnev.

Q *Who said: "We have a treaty with China, don't we?"*

A Cosmonauts Vasily Lazarev and Oleg Makarov, when it appeared that their crippled Soyuz might crash-land in China, on April 5, 1975. They landed on the Soviet side of the Altai mountains.

Q *"What we need is less crash and more program!" Who said it?*

A When a newsman suggested to Dr. Wernher von Braun that after so many failures (crashes, explosions, etc.) it appeared NASA needed a crash program, the rocket expert immediately took advantage of the opportunity for wordplay.

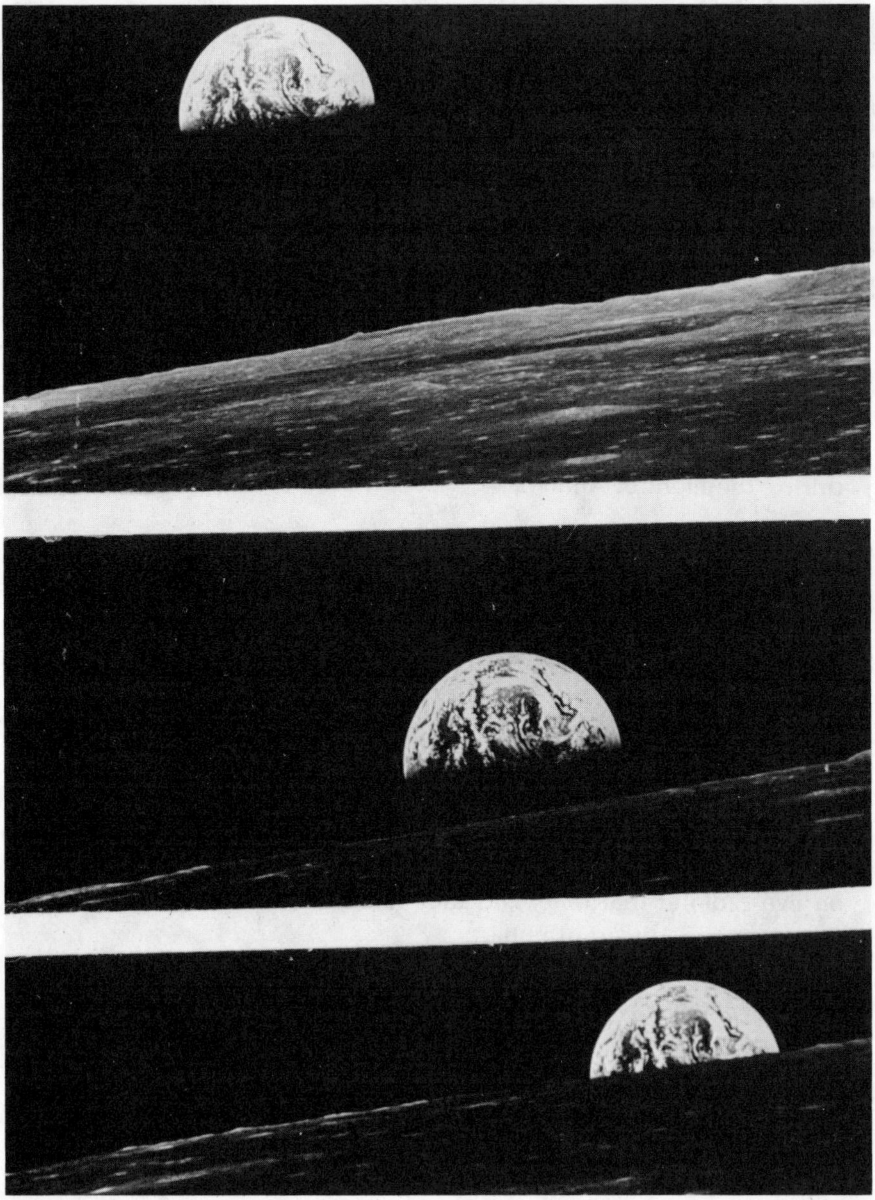

Q *Who said: "We just saw an earthrise . . . magnificent!"?*

A The crew of Apollo 10, Thomas Stafford, John Young, and Gene Cernan. They were the second humans to see the Earth as it gradually appeared over the Moon's surface, as they orbited around the Moon. This three-frame sequence was taken from the lunar excursion module during that mission, May 18–26, 1969.

Q *"Is it inhabited?" Who said it?*

A The question became a running joke among the crew of Apollo 8—Frank Borman, Jim Lovell, and Bill Anders—as they looked back at Earth from the far reaches of space.

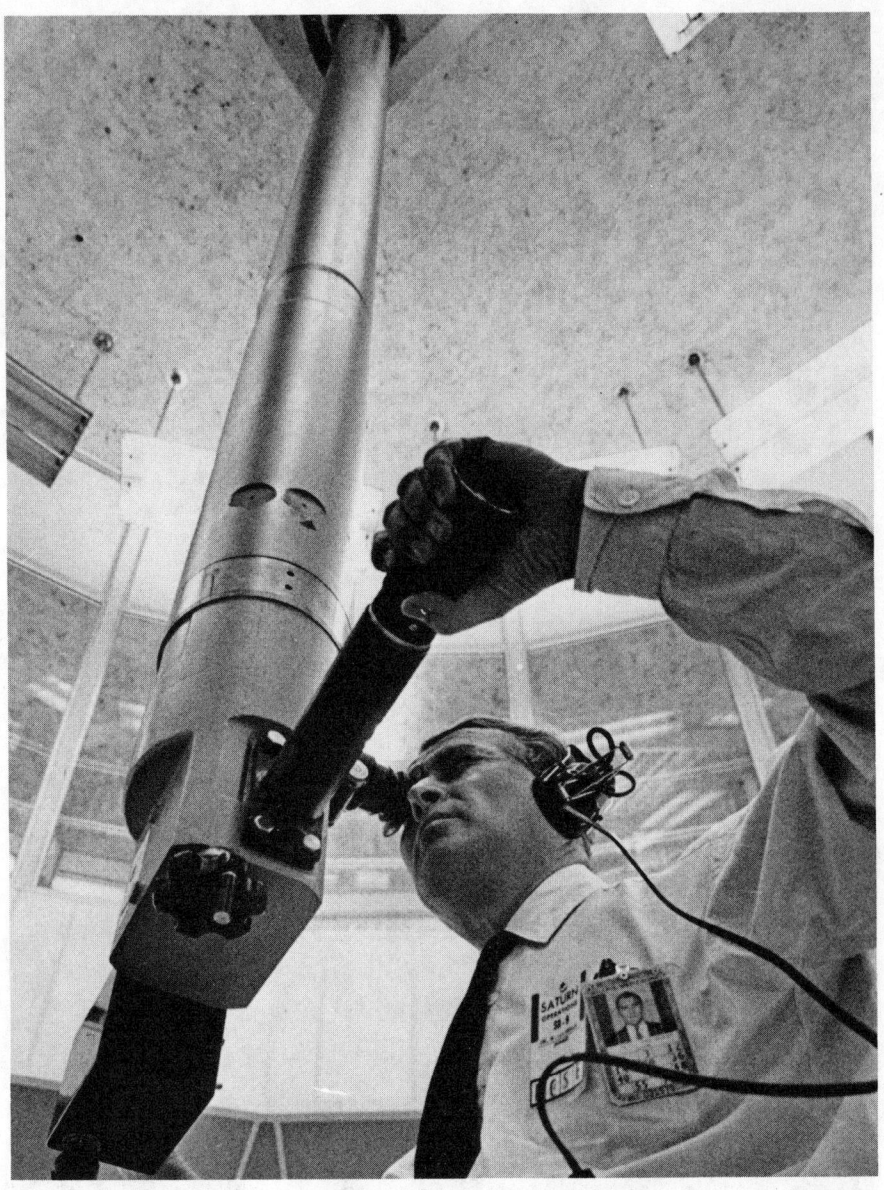

Q *Dr. Wernher von Braun was asked: "What will be the first thing we find on the Moon?" What was his answer?*

A "At this rate, an empty vodka bottle." The exchange is said to have taken place shortly after one of the several early Soviet successes in space and while the U.S. was still playing catch-up. Here, von Braun surveys the launch pad through a periscope in the Atlas blockhouse at Cape Canaveral.

> **Leave a Wake-Up Call** Despite dire predictions that the crush of Apollo 11 launch spectators would deplete local supplies of food, gasoline, and fresh water, the only real shortage was in alarm clocks.

Q *Who said: "We are in a strategic space race with the Russians, and we have been losing. If a man orbits the earth this year, his name will be Ivan."*

A President John F. Kennedy.

Q *"If you make a booster that's very powerful, I will cluster it and make it bigger." Who said it?*

A Dr. Wernher von Braun, who did just that after being brought to the U.S. after World War II. His first efforts, however, were clustering captured German V-2 rockets.

Q *Who said: "Proud as we may be of our astronauts . . . this racing to the Moon, unavoidably wasting vast sums and deepening our debt, is the wrong way [to arouse the American competitive spirit]"?*

A Former President Dwight D. Eisenhower, commenting on what he termed "stunts and unnecessary contests" in NASA's manned spaceflight program during the late days of Project Mercury.

Q *Who said: "Crash programs will fail, because they are based on the theory that with nine women pregnant, you can get a baby in one month"?*

A Dr. Wernher von Braun, in defending NASA's existing programs for achieving supremacy in space rather than giving in to the cries for "panic" programs. The remarks were made at a time when the U.S. was still considered far behind the Soviets in the space race.

Q *"Early to bed, early to rise, work like hell, and advertise!" Who is credited with saying this?*

A Dr. Wernher von Braun, when asked by a newsman to briefly define what it took to successfully conduct a space program.

Q *"If we go on producing paper at this rate, we can stack it and walk to the Moon." Who said it?*

A Dr. Wernher von Braun, commenting on the fantastic amount of paperwork related to the space program.

Gideons in Space   When a Japanese journalist asked for the text of Apollo 8's Bible reading, NASA officials told him to look in the desk of his hotel room for a large black book, then gave him the chapter and verse.

Q *Where does the following appear: "Research is reading two obscure books in order to write a third obscure book which won't be read either, but which will be used in research for a fourth book"?*

A In the NASA archives. Here Charles Duke catches up on some reading prior to the flight of Apollo 16, while the pressure in his spacesuit is being checked by technicians (not shown).

> **Creative Mishearing** Reporting on the rapidly changing space program is a difficult task even for the usually excellent Cocoa, Florida newspaper *Today*. New reporters frequently slip up. One referred to author Martin Caidin's sci-fi classic *Cyborg* as *Sideboard;* another called the Centaur rocket a "Suntower."

Q *Who said: "There has been a great deal said about a 3,000-mile high-angle rocket. In my view, such a thing is impossible today, and will be for many years to come."*

A Dr. Vannevar Bush, testifying before Congress in 1946.

Q *"One small ball in the air, something that does not raise my apprehensions, not one iota." Who said it?*

A President Dwight D. Eisenhower, commenting on the successful launch and orbiting of the Soviet Sputnik.

Q *Who quipped: "One thing I can promise you about our program . . . your tax dollars will now go farther"?*

A The witty remark was made by Dr. Wernher von Braun.

Q *Who said: "We were up to our asses in used tissues."*

A Walt Cunningham, of the cold-plagued Apollo 7, also known as the "Ten-Day Cold Capsule." The crew was concerned that they might run out of tissues before the end of the mission.

Q *Who yelled, "Let's go!" during lift-off?*

A Yuri Gagarin. *Poyekhali* became the Russian equivalent of "A-OK" in popularity.

Q *Identify the U.S. daily newspaper that went to press too early and splashed the following headline in error:* COLUMBIA DOES ENCORE— A PERFECT LAUNCH FOR "USED" SHUTTLE.

A The *Detroit News,* which announced the successful launch of STS-2. The launch was scrubbed and did not occur until two days later.

Q *Who said: "I've been howling in the wilderness for years"?*

A Gus Grissom, about poor workmanship on the Apollo spacecraft. A friend had asked him why he didn't make his concerns known to NASA management.

Q *Who said: "The flight of Valentina Tereshkova is, consequently, symbolic of the emancipation of the Communist woman. It symbolizes to Russian women that they actively share in the glory of conquering space"?*

A Capitalist, former U.S. congresswoman, and former ambassador Clare Boothe Luce in a *Life* magazine article after Tereshkova's historic flight in Vostok 6. That was before the U.S. began accepting female astronaut candidates for the second time. Seen in this October 1978 photo are, from the left: Sally Ride, Judith Resnik, Anna Fisher, Kathryn Sullivan, and Rhea Seddon.

Not Everyone Was Awed at Meeting John Glenn Young Caroline Kennedy, daughter of the President, had been promised a meeting with a space traveler. The animal-loving child, who had a well-publicized pony named Macaroni, shook hands with John Glenn, then looked around disappointedly and asked, "Where's the monkey?"

Q *Who said: "We are reserving 110 pounds of payload for recreational equipment"?*

A Mercury project director Bob Gilruth, when asked if NASA had any plans to use women in spaceflights. The crack generated considerable furor not only among feminist groups but among the media as well.

Q *"The early astronauts were the explorers. We are the homesteaders." Who said it?*

A Mary Cleave, a member of the Group 9 astronauts.

Q *Identify the speaker and location: "We have an apron all ready for you, Svetlana."*

A Valentin Lebedev spoke these words of welcome to Russia's second woman in space, Svetlana Savitskaya, as she prepared to come aboard the Salyut 7 space station, August 20, 1982.

Q *Who said: "We're really hauling . . . the mail"?*

A Gus Grissom, as Gemini 3 lifted off.

Q *Which spacecraft was named by the CapCom?*

A Gemini 3. As Grissom and Young made the first U.S. two-man flight, Gordon Cooper called out, "You're on your way, *Molly Brown*," confirming the name they had been told not to use.

Q *With what statement did Alan Shepard begin his postflight debriefing in 1961?*

A "My name José Jimenez, Chief Astronaut, Junited States."

Q *Who said: "That's one hell of a headache, but a short one"?*

A John Young, watching a test of the Gemini ejection seat. The seat worked just fine, but the hatch failed to open; the seat blasted right through the 2-inch-thick hull. (Young had other problems with ejection seats; during an early shuttle test, it was determined that the parachutes would open "about 50 feet after we hit the ground.")

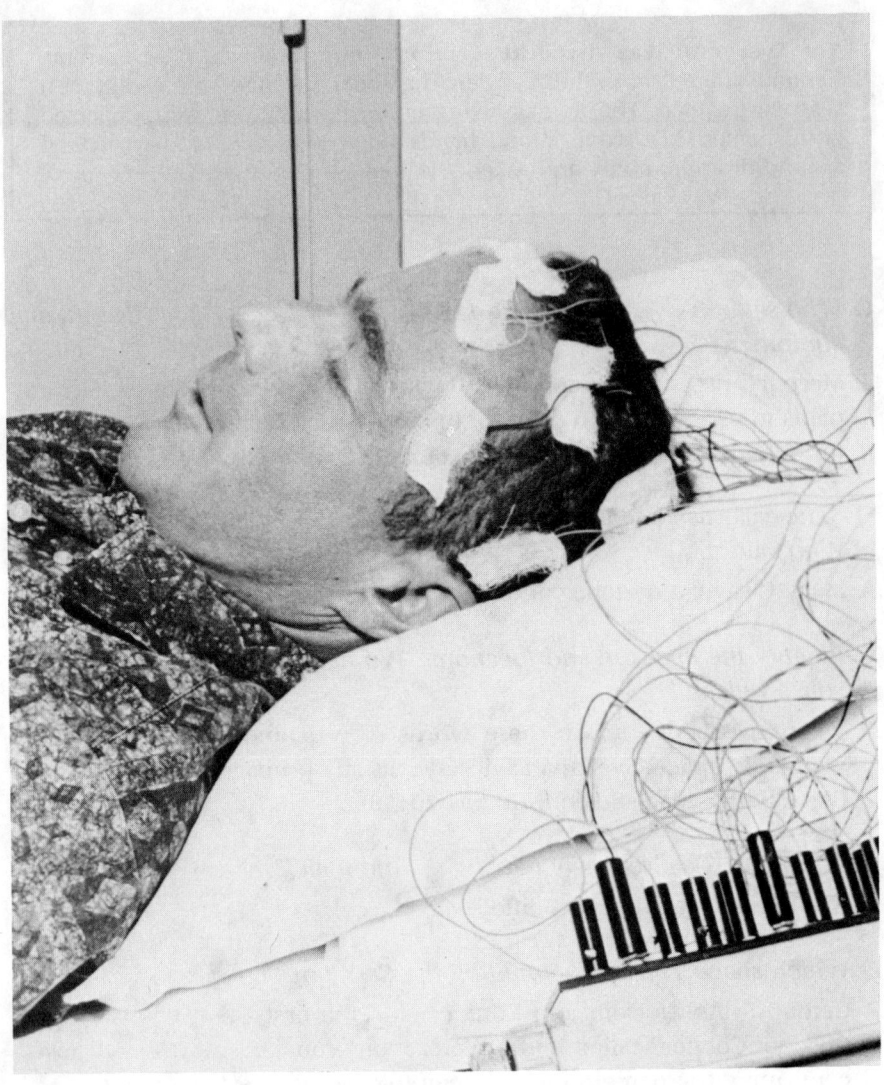

Q *Identify the astronaut who described the first sunset he saw in space as: "Spectacular, like a very brilliant rainbow"?*

A Malcolm Scott Carpenter during his singular trip in space, *Aurora 7*, in the Project Mercury program. The flight was originally set to be flown by Deke Slayton but Carpenter moved up when Slayton was grounded by a heart irregularity. As a result Carpenter became the second American to orbit Earth. He was also able to determine that the "fireflies" John Glenn saw were frost particles and photographed them. Carpenter is seen here undergoing a preflight electroencephalograph during a 5-hour physical.

> **Star Talk?**   President Ronald Reagan, a former Hollywood movie star, closed his good-luck letter to the STS-7 shuttle crew with the phrase "May the Force be with them."

Q *The following headline appeared in a Milan, Italy, newspaper: TOO PERFECT—THE PUBLIC IS GETTING BORED. What was it referring to?*

A The mission that came the closest to ending in disaster, Apollo 13.

Q *Who said: "The spacecraft is in real good shape as far as we are concerned. We're bored to tears down here."*

A Capsule communicator Joe Kerwin, less than an hour before Apollo 13 sustained oxygen tank explosions.

Q *"Houston, we've got a problem here." Who said it?*

A The message was radioed by Apollo 13 after two onboard explosions rocked the ship. The mission was aborted following a daring gravity loop around the Moon.

Q *Who said: "I'm afraid this is going to be the last Moon mission for a long time"?*

A Astronaut Jim Lovell, during the return leg of Apollo 13.

Q *What did the celluloid buttons reading HORNET PLUS THREE signify?*

A They were made to commemorate the recovery of the Apollo 11 crew by the aircraft carrier USS *Hornet,* after the historic Moon landing mission.

Q *"If you can't take any better care of a spacecraft than this, we might not give you another one." Who said it?*

A CapCom Joe Kerwin to astronaut Jim Lovell, just before the re-entry of Apollo 13.

Q *Which astronaut said: "Get this fucking hatch open" during a dangerous moment after an Apollo splashdown?*

A Tom Stafford, as lethal nitrogen tetroxide gas filled the spacecraft during re-entry from the Apollo-Soyuz mission. An intake valve had not been closed properly. The world didn't hear most of the comments because of a weak radio link. Vance Brand was knocked unconscious; all three had their lungs blistered.

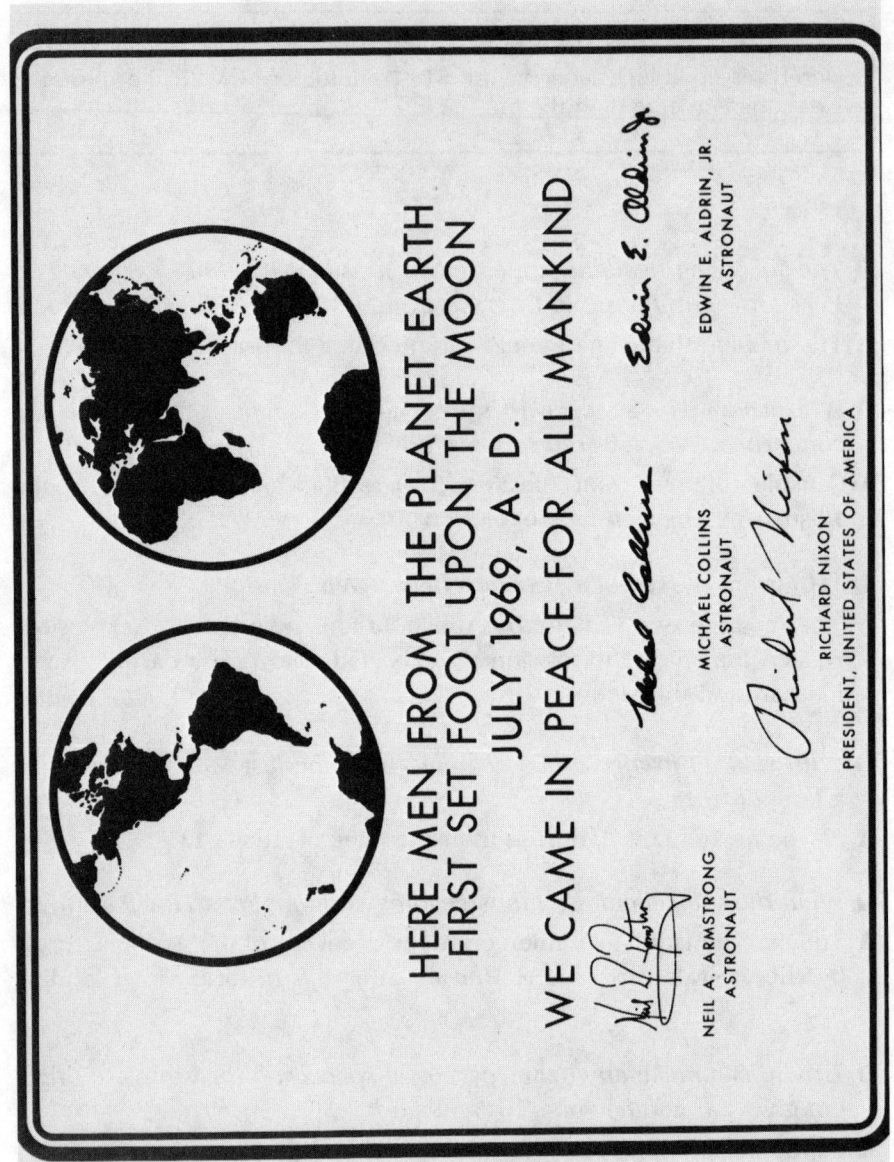

Q *"We came in peace for all mankind." Who said it?*

A It is the inscription on a plaque that was attached to the first spacecraft flown to the surface of the Moon, the lunar excursion module named *Eagle*, by the crew of Apollo 11. The plaque is fastened between the third and fourth rungs of the ladder on the LM's descent stage, which remained behind.

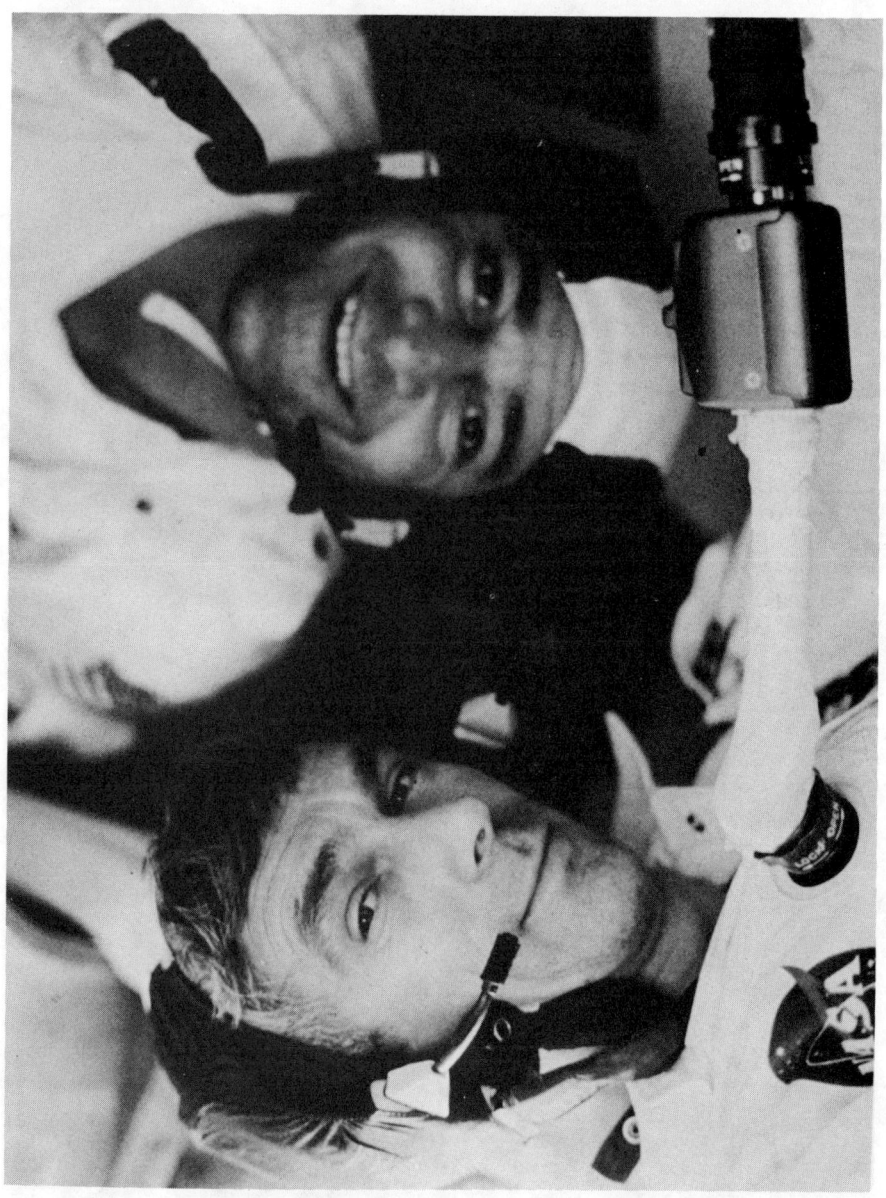

Q *What were the last words spoken on the surface of the Moon?*

A "We leave now as we once came, with peace and with hope for all mankind," spoken by Gene Cernan. Unofficial sources say the *real* last words were "Let's get this mother out of here." Cernan (left) and a weightless Ronald Evans took time for a farewell photo as Apollo 17 prepared to leave lunar orbit.

Coincidence or Prediction? Interesting parallels between Jules Verne's fictional nineteenth-century account of a manned flight to the Moon and the Apollo manned flights include:
- Both had three-man crews.
- Flights in both blasted off from bases in central Florida.
- Astronauts in both were secured in couches at launch.
- Verne's craft was *Columbiad*, Apollo 11 was *Columbia*.
- *Columbiad* had a crewman named Ardan, *Columbia* had Aldrin.
- Both were the result of military technology in a postwar era.
- Both used retro-rockets for descent.
- Both landed on water rather than land, in the Pacific.
- Payload dimension was approximately the same.
- Florida and Texas sought control of both missions.
- Air in both crafts was recycled through potash to absorb CO_2.

Q *Whom did Neil Armstrong call "my most competent critic"?*

A Buzz Aldrin, his co-Moonwalker on the historic Apollo 11 flight. Armstrong (center), Aldrin (right), and third crew member, Michael Collins, are seen here reunited during ceremonies in Washington in July 1979 to observe the tenth anniversary of their Moon landing.

Q *Who said: "I'm sick of steak."*

A Neil Armstrong, two days (and two steaks) before launch.

Q *Who described the Moon this way:*

1. "Magnificent desolation."

A Buzz Aldrin, during Apollo 11.

2. "A very dark and unappetizing place."

A William Anders, during Apollo 8.

3. "It has a stark beauty all its own; it's much like the high desert of the United States."

A Neil Armstrong, during Apollo 11. The comparison is to the area around Edwards Air Force Base.

Clergy Disagree After the Apollo 11 spacecraft returned from the Moon and landed in the Pacific, the astronauts were kept in quarantine aboard the carrier *USS Hornet*. President Richard Nixon went out to greet them, speaking over a phone and looking at them through a small porthole. He said, "This is the greatest week since the Creation." Billy Graham and other Christian leaders were offended, suggesting that the Crucifixion might still outweigh the lunar landing.

Q *What respected publication, commenting on Dr. Robert Goddard's research, stated that a rocket would not function in the vacuum of space?*

A The *New York Times*, in 1920. The article also said Goddard "only seems to lack the knowledge ladled out daily in our high schools." A retraction was printed as Apollo 11 sped to the Moon. But the *Times* has made other faux pas in covering science and technology: it predicted the Brooklyn Bridge couldn't handle rail traffic, discounted electrical power and the incandescent bulb, predicted the airplane would never fly, and opposed the artificial heart.

Q *Who described the Earth as "a grand oasis in the great vastness of space"?*

A James Lovell, during the Apollo 13 mission.

Q *What sign was tacked to the quarantine trailer as the astronauts returned to the* Hornet?

A Please Do Not Feed the Animals.

Q *Who said: "Things are beginning to stack up a little"?*

A Gordon Cooper, aboard *Faith 7*. The understatement came after several problems arose: a false signal that automated re-entry had begun ahead of schedule, the loss of all altitude readings, and the complete loss of power in the automated control system, which forced Cooper to make the first completely manual re-entry.

> **Worker's Paradise *Really* Means . . .** There was confusion over just how Yuri Gagarin landed, when Soviet officials reported that he parachuted to the ground and landed "on both feet, without tumbling." Other reports, including Gagarin's own, said he had stayed with the Vostok all the way down. Soviet officials claimed the Russian expression "on both feet" is the pilots' equivalent of the American expression "three-point landing." Despite this subterfuge, other cosmonauts later confirmed that Gagarin *had* ejected.

Q *Who said: "NASA can't afford to lose astronauts; if they lose astronauts they lose funding."*

A Sally Ride, discussing the safety of spaceflight.

Q *"I guess they want astronauts to be tiddlywinks players." Who said it and why?*

A Gordon Cooper. The night before he was to race at Daytona, NASA forbade it. Cooper was in a racing team along with Gus Grissom and Charles Buckley, the Kennedy Space Center security chief. NASA later amended the rule to forbid racing only by astronauts with upcoming flights.

Q *Identify the mission that prompted this outburst: "We've discovered pizza!"*

A Voyager 1. The statement was made by a technician at Pasadena's Jet Propulsion Laboratory in 1979 after the deep space probe transmitted photos of the splotchy red, yellow, and white surface of the Jovian moon Io.

Q *What American newspaper columnists predicted the launch of Sputnik 1?*

A Drew Pearson and Jack Anderson, in their "Washington Merry-Go-Round" in 1956. When the Soviets announced their intention to launch a satellite at least a year beforehand, most Western journalists ignored or discounted the reports.

Q *Who was the subject of the film* How to Succeed in Space Without Really Flying Much?

A Alan Shepard, at a 1967 roast celebrating the sixth anniversary of his 15-minute flight. In the interim, he had parlayed his fame into a powerful position inside NASA and a lucrative business and banking career on the outside.

Q *Who said: "Can't you fellows invent some other race here on Earth that will do some good?"*

A President John Kennedy, speaking to science advisor Jerome Wiesner in 1962, before declaring the national goal of going to the Moon.

Q *Who said: "If it hadn't worked, they'd be asking for your head"?*

A James E. Webb, NASA administrator, to Robert Gilruth, director of Project Mercury, after the successful launch of Alan Shepard. The comment was made while the men were riding in an open limousine during Shepard's motorcade in Washington, D.C., and saw the enthusiasm of the crowd. Later that day Webb received President John F. Kennedy's congratulations on behalf of everyone involved in NASA.

> **We Can Do It Ourselves, Thank You**　Members of Congress were so afraid that President Kennedy might show interest in a joint U.S.-U.S.S.R. Moon landing that they attached this rider to the bill authorizing NASA's 1964 budget: "No part of any appropriation made available to NASA by this Act shall be used for expenses of participating in a manned lunar landing to be carried out jointly by the United States and any other country without consent of the Congress."

Q *Who said: "It's unfortunate, but the way the American people are, now that they have developed all of this capability, instead of taking advantage of it they'll probably just piss it all away"?*

A The candid remark was made by President Lyndon Johnson while touring a space contractor facility with a group of astronauts. He was expressing concern over public opinion following the fatal fire on January 27, 1967, which claimed the lives of three astronauts and generated, in some quarters, calls to abandon the manned spaceflight program to reach the Moon.

Q *Who said: "The space age devours people. I have been completely devoured by this business"?*

A Wally Schirra. It was his explanation for resigning from NASA instead of seeking a lunar mission.

Q *"It makes me proud to be part of a country that can send a man 250,000 miles away from home, and even prouder to be part of a country that can bring him back." Who said it?*

A Ken Mattingly, after his return from the Apollo 16 lunar mission, April 1972.

Q *"Do you realize what we have accomplished today? Today the spaceship was born!" Who said it and why?*

A Gen. Walter Dornberger said it to Wernher von Braun after the first successful V-2 rocket launch, October 3, 1942.

> **The Ides of March March On**　After years of delays and setbacks, NASA announced that the shuttle would definitely fly in March 1981. Kennedy Space Center workers printed up calendars showing *every* month after February labeled "March."

Q *Who asked Deke Slayton the following question from space, before a television audience estimated at several million: "Are you a Turtle?"*

A Prankster Wally Schirra. In military circles the Turtles are a fun-loving group of personnel who frequently ask the question of each other for this reason: failure to reply requires the silent Turtle to buy a round of drinks for all who heard the question. The reply is "You bet your sweet ass I am," and Slayton went on a closed-circuit hookup to answer Schirra's challenge. Schirra had done the same when Slayton challenged him during Mercury 8 launch preparations.

> **Thanks Guys, We Needed That**   After their return from the Moon, Apollo 12 astronauts Conrad, Bean, and Gordon went on an international goodwill tour and were perplexed at the small turnout for a parade in Iran. They discovered that local communists had printed leaflets showing a parade route several blocks away from the actual one.

Q *Who were the Turtles, and who was their leader?*

A Dreamed up by test pilots during World War II, the Interstellar Association of Turtles believes that you never get anywhere in life without sticking your neck out. Every Turtle presumedly owns a jackass, and when asked "Are you a Turtle?" he must answer, "You bet your sweet ass I am," or buy a round of drinks. Wally Schirra was High Potentate of the Turtles, Outer Shell Division.

Q *Who sent this message from orbit?: "I have my four hours' required flight time for the month and request that a flight chit be established for me."*

A John Glenn sent that message to the U.S. Marine Corps commandant during his 1962 flight.

Q *Name this speaker: "Tossing a man up in the air and letting him come back . . . has about the same technical value as the circus stunt of shooting a young lady from a cannon."*

A Dr. Hugh Dryden, director of the National Advisory Council on Aeronautics, criticizing the Army's Project Adam suborbital flight plan in a congressional hearing. A few years later, as the number-two man at NASA, he supervised exactly the same "circus stunt."

Q *Identify the speaker and the mission: "The fact that I could take over and show that a pilot can control the capsule manually satisfied me the most."*

A John Glenn, after the flight of *Friendship 7* (MA-6), on February 20, 1962.

> **Ask and You Shall Receive, Comrade**   When U.S. astronauts went to Russia to train for the Apollo-Soyuz test project, they presumed their rooms were bugged. One astronaut complained to the walls that there weren't enough coat hangers; when he came back after dinner, the room was strewn with them. He gratefully thanked the lampshade.

Q *How did John Young sum up America's return to space on the first shuttle mission?*

A "The dream is alive again."

Q *Who left a plaque with the following inscription on the Moon: "Here man completed his first exploration of the Moon, December, 1972 A.D. May the spirit of peace in which we came be reflected in the lives of all mankind"?*

A The crew of Apollo 17, the last lunar mission in the Apollo program. It is signed by Harrison Schmitt, Ron Evans, Gene Cernan, and President Richard M. Nixon. The crew got together for this photo with their lunar rover some four months before the flight in 1972.

> **From Earth with Love** A tiny silicon disk left on the Moon by the Apollo 11 crew contained messages from seventy-three heads of state, as well as former U.S. presidents Eisenhower, Kennedy, and Johnson, who had all presided over NASA.

Q *"What are they going to do? Shoot you up in the nosecone of an Atlas?" Who said it to whom?*

A Betty Grissom said it to her husband, Gus, after he received sealed orders to report to a Washington hotel in civilian clothing. Betty came up with the most ridiculous scenario imaginable; the Atlas had been the Air Force's most public failure for months. Little did either suspect that that was exactly what the government had in mind in astronaut selection ... finding pilots to ride a Mercury capsule atop an Atlas rocket.

Q *"If we don't make this next maneuver correctly, you won't get your pictures developed." Who said it?*

A Jim Lovell made the comment to Fred Haise and Jack Swigert, as they prepared for a rocket burn to bring their crippled Apollo 13 around the Moon and back to Earth. Any error would have lost them in space forever. Haise and Swigert had been photographing the far side of the Moon.

Q *What soul-stirring command sent men out of Earth orbit and toward the Moon for the first time?*

A "You are go for TLI [trans-lunar injection]." The equally thrilling response for this momentous occasion was "Roger, Houston." Ironically, the CapCom who gave the historic command was writer and poet Mike Collins, who decried the lack of colorful phrases in the program. Seems he blew his chance.

Q *Name the speaker and the subject: "It's like the old fighter pilot's life ... long periods of boredom punctuated by moments of stark terror."*

A Astronaut Dick Gordon concerning lunar flights.

Q *What was ironic about this news release? "Komarov is in good health and feeling well."*

A Vladimir Komarov's Soyuz 1 spacecraft malfunctioned during re-entry, killing him on April 23, 1967. The release was premature.

Q *Name the scientist called "Moon Man" by his detractors.*

A Dr. Robert H. Goddard, who gained the nickname because of his fascination with the relationship between rockets and space travel. This 1916 photo shows him with a circular vacuum tube he used to prove that there was greater rocket efficiency in a vacuum than in air.

Dreamers and Pioneers

Q *Name the original six pilots selected to fly the proposed X-20 rocket plane, which was never built.*

A Selected on September 19, 1962, they were: Maj. James W. Wood, Maj. Russell L. Rogers, Maj. Henry C. Gordon, Capt. William J. Knight, Capt. Albert H. Crews, all U.S. Air Force officers, and civilian test pilot Milton O. Thompson of NASA. The X-20 would have had 2.5 million pounds of thrust, compared to the 367,000-pound maximum produced by Project Mercury's Atlas rockets. It would have been the first ship to go into orbit with a pilot at the controls from beginning to end.

Q *Who was the first man to break the sound barrier?*

A Col. Chuck Yeager, on October 14, 1947, at Edwards Air Force Base, California, while testing the Bell Aircraft X-1 jet plane. The Air Force kept the event secret until June 1948.

Q *Who was the first man to fly a rocket plane?*

A Capt. Homer Boushey of the U.S. Army Air Force, in 1940. The experiment consisted of a lightweight Ercoupe airplane with six rockets, fueled by bottled liquid explosives, attached to the wings. Immediately upon ignition the aircraft was thrust into flight. The procedure is called JATO (jet-assisted takeoff).

Q *What was Operation Paperclip?*

A The U.S. Army's secret project using German scientists to test-launch captured V-2 rockets.

Q *Who designed and tested the first Soviet liquid-fuel rocket engine?*

A Friedrich Arturovich Tsander, a Latvian engineer who carried on the work of Tsiolkovsky. He died about five months before the first test flight of the rocket.

Q *What space pioneer and NASA consultant was forbidden to read his own work?*

A Hermann Oberth, whose research in the 1920s paved the way for manned spaceflight. Born in Transylvania, he was a German citizen during World War II. Coming to NASA in the 1960s, his work was classified as soon as it was turned in; as an alien, he couldn't get security clearance to see it, even for reference. Oberth (foreground) joined, clockwise, Ernst Stuhlinger, Gen. H. N. Toftoy, Eberhard Rees, and Wernher von Braun in Huntsville, Alabama, for this 1956 photo. Toftoy was the creator of Operation Paperclip, the operation that brought Germany's rocket experts to the U.S. after World War II. The rest were part of the U.S. Army's Ballistic Missile Agency at the time.

1933 World's Fair Lit from Space   The organizers of Chicago's 1933 World's Fair decided that something special was needed for it to live up to its theme, "A Century of Progress." As a result, they amplified rays from the star Arcturus (which were received by a series of telescopes) and used this power to turn on the lights each night. The power from Arcturus, 40 light-years away, had left the star in 1893, the year of the first Chicago World's Fair.

Q *Name the first foreign target to be hit by an American rocket.*

A A cemetery outside Juarez, Mexico. While German scientists were stationed at Ft. Bliss, Texas, under Operation Paperclip, a captured German V-2 rocket was launched into uninhabited desert. It followed instructions perfectly; unfortunately, the guidance gyroscope had been installed backwards. Apologies were made to Mexico, and according to Dr. Ernst Steinhoff, when American officials reached the scene, some young Mexicans had set up a stand and were selling rocket fragments to tourists.

Q *Who wrote the first work of science fiction?*

A The first clearly fictitious work is *A True History* by Lucian of Samosata, written in 165 A.D. In it, he travels past Gibraltar and is caught in a tornado which dumps him on the Moon, where he finds the humanoid Lunarians at war with the Solarians over the rights to colonize the "Morning Star."

Q *How did the Soviets reward their military rocket research team in the 1930s?*

A By throwing them in prison. Stalin believed that armaments minister Mikhail Tukhachevsky was plotting against him, and in 1937 he and his senior staff were executed. One-star general Sergei Korolyov and junior staffers were simply shipped to Siberia. Those who survived were brought back after hostilities began with Germany.

Q *Who originally headed the German rocket program?*

A Walter Dornberger, who enlisted the aid of the young Wernher von Braun. Ironically, both men said after the war that rockets are expensive and inefficient weapons, and that they were sold to the military only as a way to get research funding. Robert Goddard and Sergei Korolyov took the same route with the U.S. and Soviet governments.

Q *Who was the first science-fiction writer to use rockets to get his hero to the Moon?*

A Cyrano de Bergerac, in *Voyages to the Moon and Sun* (1657). Other writers used cannon, balloons, flapping wings, antigravity, or nontechnological means. The illustration is from the first edition of that work.

The Sun on the Moon   Mankind's interest in the Moon and the possibility of life on it spans the centuries. However, what may have been the first newspaper story dealing with the subject appeared in the *New York Sun* in 1835. Reporter Richard A. Locke dreamed up a story about a "vast population" of near-human beings living on the Moon and then ascribed it to astronomer Sir John Herschel. The story received widespread circulation before being denounced as a hoax.

Q *Who was the director of America's first serious effort to contact life in space?*

A Dr. Frank D. Drake, the father of Project Ozma. He has computed that there may be 10 million intelligent societies in our solar system alone. A radio astronomer, Drake began a series of radio signal hunting in 1959 with the expectation that it would take eleven years for a signal to be reached on Earth. The two stars Drake concentrated his signals on are Tau Ceti and Epsilon Eridani.

Q *Who was the first working scientist to write science fiction?*

A Johannes Kepler, who also discovered the three laws of planetary motion. In *Somnium* (1632), Kepler's hero reaches the Moon in his sleep.

Q *Name the first English work of science fiction.*

A *The Man in the Moone*, published posthumously by Bishop Francis Goodwin in 1638. The hero is not from England but from the center of contemporary exploration, Spain.

Q *Which scientist devised a complicated equation on the side of a streetcar?*

A Dr. Hugh L. Dryden, who used the first available surface to illustrate a point for a colleague. When the trolley began moving again, Dryden chased it through much of New Orleans' French Quarter.

Q *Who imprisoned Wernher von Braun during World War II, and why?*

A Reichsführer Heinrich Himmler, chief of the Gestapo, imprisoned von Braun for two weeks at Stettin Prison on the suspicion that the scientist was interested in building rockets only for spaceflight, not for warfare. It took Gen. Dornberger's personal appeal to Hitler to spring von Braun.

Q *Where is the first suggestion that the lunar surface may have less gravity than here on Earth?*

A In the first English work of science fiction, Francis Goodwin's *The Man in the Moone.* His hero, Domingo Gonsales, also finds the Moon to be mostly covered by water. Another English work, *Discovery of a New World in the Moone* by John Wilkins (1638), explains that the Earth is shrouded by an "orb of magnetic vigor" which dissipates at 20,000 miles altitude. Both works predate Newton's discovery of the laws of gravity and motion. Photo above was taken by Apollo 14 in February 1971. It shows the Davey Crater Chain.

Q *Name the first piece of sculpture left on the Moon.*

A *The Fallen Astronaut,* a tiny aluminum figure by Paul Van Hoeydonck, placed in a lunar crater on August 2, 1971, by Apollo 15 crewmen Dave Scott and Jim Irwin. They agreed to leave it as a tribute to fallen comrades, with the understanding that Van Hoeydonck wouldn't capitalize on it. He later offered 950 copies of the work for $750 each. The figure is seen resting before a plaque that contains the names of fourteen American and Soviet astronauts who died in the race to the Moon.

Peaceful Prediction   An eclipse was once responsible for preventing a war. The Greeks, who could calculate eclipses mathematically, warned the hostile Lydians and Medes that as a sign of their power they would darken the Sun by making the Moon pass in front of it on a given day—May 28, 585 B.C. When the Lydians and Medes witnessed the eclipse on that day, they were so frightened that they canceled their plans to attack Greece.

Q *Who headed the task force that proposed the space shuttle?*

A Vice President Spiro T. Agnew, seen here in conversation with Wernher von Braun at Kennedy Space Center in 1972.

Mother Knew Best   When Wernher von Braun was baptized at the age of 8, his mother broke family tradition by giving him a telescope instead of a gold pocket watch.

Q *How was Peenemünde chosen as the German missile development site?*

A Von Braun's mother suggested the isolated, marshy Baltic seacoast since it had been the elder von Braun's favorite duck-hunting area.

Q *What is the first mention of retro-rockets in science fiction?*

A In "The Unparalleled Adventure of One Hans Pfall," by Edgar Allan Poe, 1835, the protagonist uses gunpowder blasts to slow his descent to the Moon—not in a rocket ship, but in a balloon. The concept of a device to slow descent is even older. In 1705 a Moon traveler in a machine with steam-powered flapping wings reversed their pitch as he approached the Moon—in Daniel Defoe's "The Consolidator, or Memoirs of Sundry Transactions from the World of the Moon."

Q *What was the X-20 project called?*

A Dyna-Soar, short for Dynamic Soaring. The large delta-winged craft would have been launched atop a Titan rocket and would have glided to a runway landing, like its direct descendant, the space shuttle. The plans were scrubbed when NASA decided to use the ballistic pod approach.

Q *Who designed the Dyna-Soar?*

A Walter Dornberger, von Braun's military boss in Germany.

Q *What was America's first proposed spacecraft?*

A The X-15-B, a vertically launched version of the famous rocket plane. It would have landed on a runway.

Q *Identify the first human to experience deceleration of a vehicle equal to thirty-five times the force of gravity and wind pressure of more than 2 tons.*

A Col. John P. Stapp, U.S. Air Force, aboard a specially built speed sled at Holloman Air Force Base, Alamogordo, New Mexico, in December 1954. He was a pioneer explorer of the limits of human endurance. His experiences with g-forces played an important part in the training of astronauts a few years later.

Q *Who was the first geologist on the Moon?*

A Jack Schmitt, of Apollo 17. He was sent after years of forceful arguments from the National Science Academy about the value of a real scientist in space. On the Moon he found an unusual rock and identified it as "only" 25 million years old. Earthside lab analysis felt he missed by "only" 3.5 billion years. Schmitt is seen here suited up for tests prior to the sixth, and final, lunar mission in Project Apollo.

> **Oberth Plan** In the book *The Rocket in Interplanetary Space*, published in 1923, scientist Hermann Oberth detailed an approach to sending and retrieving a manned lunar mission very similar to the one actually used by NASA.

Q *How did the Juarez trolley line supply scientists for the American rocket effort?*

A When the Army decided to transfer over a hundred former German rocket scientists from Texas to Huntsville, Alabama, in 1949, there was a problem with immigration laws. The Germans had been brought in illegally under Operation Paperclip, so they had to go to Mexico so they could come back into the U.S. legally. Where U.S. immigration forms ask "method of entry," the Germans wrote "Juarez Trolley Line."

Q *Why were U.S. rocket scientists upset with President Kennedy's goal of a lunar landing by 1970?*

A They thought the short-term goal of a man on the Moon would divert resources and attention from the more difficult but more promising goal of a manned landing on Mars, which had been the dream of Goddard, von Braun, and many others.

Q *Who introduced Lindbergh to Goddard?*

A One of Goddard's students and a former associate of the Wright brothers: Edwin Aldrin, Sr., father of our second moonwalker.

Q *Name the first scientist to suggest life on the Moon.*

A Anaxagoras of Greece (500–428 B.C.), who also believed that the Moon had mountains and that moonlight was reflected sunlight, and who explained the phases of the Moon, said the Sun was larger than the Peloponnesus, and claimed that the stars were "fiery stones," not gods.

Q *What was the first literary mention of null gravity points?*

A Such points, where the gravitational attraction of one body is canceled by that of another, are mentioned by Johannes Kepler in *Somnium*, 1632.

Q *When was the first Atlas launched?*

A America's first long-range weapon, the Atlas ICBM, was successfully launched on December 17, 1957.

So Close, But So Far—Christopher Kraft, NASA's flight director for all of the Project Mercury and some of the early Gemini flights, never got to attend any of their lift-offs. Because of his job, Kraft had to view these historic events from the control center via a TV monitor. Kraft is seen here at his console (right) during the recovery of Apollo 15. At this point he was deputy director of Manned Spacecraft Center, Houston.

Ward's Chinese Rockets Huntsville, Alabama, newspaperman Bob Ward was included in the first formal exchange of journalists with China in 1980. He doggedly pressed his guides to show him a Chinese aerospace facility. Finally he was taken to a Shanghai boy's club and shown a display case of handmade model rockets.

Q *What legal precedent did the Soviets set with Sputnik?*

A That orbiting spacecraft do not violate sovereign airspace. Since the Soviets didn't ask permission of the nations under Sputnik's flight path prior to launch, we didn't ask their permission to use spy satellites over their landmass.

Q *How did Charles Lindbergh help secure the future of spaceflight?*

A In 1927 he convinced philanthropist Daniel Guggenheim and the Andrew Carnegie Foundation to donate over $25,000 to the work of Dr. Robert Goddard. The endowment allowed Goddard to work on rocketry full time.

Q *Identify the novel, which eventually became a movie, that describes a fictitious Project Mercury flight designated MA-10.*

A Martin Caidin's *Marooned.* By the time it was adapted for Hollywood, the flight described had become an Apollo-type mission in which the astronaut crew experienced a problem in space and faced death.

Q *What is "terraforming"?*

A The redesign of another planet to more closely resemble the Earth, or to allow terrestrial life-forms (such as humans) to live in its environment without artificial life support.

Q *Who first conceived of terraforming Mars?*

A Konstantin Edwardvich Tsiolkovsky (1857–1935), the Russian mathematician who became "the father of spaceflight." He also believed that men would travel in space by reactive force. He proposed using two fluids, liquid hydrogen and liquid oxygen, united in combustion to create the necessary thrust to produce rocket flight.

Q *Who first determined the relative sizes of the Earth and the Moon?*

A Greek astronomer Hipparchus (190–120 B.C.), who did it with simple geometry.

Q *Which space pioneer was born in a castle?*

A Wernher von Braun, born March 23, 1912, in Wirsitz Castle, East Prussia. He was the son of Baron Magnus von Braun and Baroness Emmy von Quistorp. He is seen here (right) with Kurt Debus at Cape Kennedy.

Q *What technical ability persuaded Hitler to use the V-2 rocket?*

A The new British invention, radar, which was helping the RAF splatter incoming German bombing raids. Dornberger and von Braun convinced Hitler that the V-2's liquid-powered trajectory would carry it above the British radar shield.

Q *What was the first story of an alien attack on Earth?*

A "The War of the Worlds," by H. G. Wells, written in 1897.

Q *What did the Germans call the long-range A-10 rocket they were developing when World War II ended?*

A The Amerika rocket. According to one engineer, Ludwig Roth, although it never flew, "it got us here anyway."

Q *For whom was Edwards Air Force Base named?*

A Col. Glenn Edwards, a test pilot who died there in 1950 while testing a plane called the Flying Wing. Film of the crash was used as part of the opening for "The Six Million Dollar Man."

Q *What is the first historical mention of military rockets?*

A Twelfth-century A.D. Chinese armies used battlefield rockets; the Mongols used rockets in their sack of Baghdad in 1258.

Q *Where did the practice of the countdown originate?*

A Before World War I, the German general staff put together a plan for overwhelming the French army in the fall of 1914. Their time line was counted in days before (minus) or after (plus) mobilization. Hence, M − 1 was the day before mobilization, M + 5 was five days after. The first launch countdown occurred in a German science-fiction film of the period.

Q *Who made the first serious proposal for manned space flight?*

A German engineer Hermann Gaswindt, in 1891.

Q *Identify the first U.S. satellite successfully launched.*

A Explorer 1, launched on a U.S. Army Jupiter C rocket on January 31, 1958. The 6-foot cylindrical Explorer satellite went into an orbit with an apogee of 1,587 miles and perigee of 219 miles. It weighed 31 pounds, approximately one thirty-sixth the size of the Soviet Sputnik 2.

Unmanned Space Probes

Q *When did the first man-made object land on another celestial body? What was it?*

A On September 13, 1959, the Soviet Union's Luna 2 crashed on the Moon intentionally.

Q *When is the interplanetary age considered to have been born?*

A December 14, 1962, when the Mariner 2 spacecraft successfully completed a flyby of the planet Venus. It was the first time man had conquered the depths of space to the vicinity of another planet. The 109-day trip covered 180.2 million miles.

Q *When did the U.S. attempt to launch its first Vanguard satellite?*

A On December 6, 1957. Unfortunately the 72-foot rocket and its 3¼-pound satellite packed with electronic gear rose only 5 feet off the launchpad at Cape Canaveral before exploding in flames.

Q *Name the shortest flight of an American spacecraft.*

A Mercury-Redstone 1, also known as the "Popped Cork." The booster shut down after rising only 4 inches, enough to count as a lift-off.

Q *Name the first flight of a robot supply ship.*

A *Progress 1* left the U.S.S.R. on January 20, 1978, with food and other supplies for the two-man crew of Salyut 6.

Q *Which is brighter, the Earth or the Moon?*

A The Earth, which reflects 38% of the sunlight it receives. Astronomers express this as "Albedo 0.38." The Moon reflects only 7%.

But Can They Get MTV?   The first TV broadcast has now traveled far enough to reach more than 400 stars.

Q *Who said "Millions for chimps, but not one cent for women"?*
A Jerrie Cobb, a female astronaut trainee, on the elmination of the women's astronaut candidate testing program in July 1961. However, female chimp Glenda didn't seem disturbed by the turn of events as she continued training for the Mercury-Redstone unmanned (or un-womaned) series.

Bonzo's Classmates? NASA established the Aeromedical Research Laboratory at Holloman Air Force Base, New Mexico, in 1959, to train chimpanzees for spaceflight. Nicknamed "Chimp College," the facility housed forty chimpanzees who were put through physical and psychological tests. They were trained to do fairly complex manual tests under threat of operant conditioning: psychomotor stimulus plates were attached to the chimps' feet, and if they failed to correctly perform a task they received an electric shock. The chimps began training for spaceflight at the same time as the Original 7 Astronauts. By the time a chimp was selected for an actual flight, they had gone through all phases of astronaut training except abort and re-entry emergency sequences and attitude control of a spacecraft in flight.

Q *Which space traveler was pulled from a mission because of religious conflicts?*

A A fully trained rhesus monkey. A native of India, he was replaced on the May 28, 1959, launch of a Jupiter rocket by a partially trained but American-born rhesus called Able, because President Eisenhower feared offending Indians, who consider the rhesus a sacred animal.

Q *How long did Enos live after his orbital flight?*

A The first chimpanzee to orbit the Earth died on November 4, 1962, ten months after his 2-orbit flight. Cause of death was shigellosis and dysentery, completely unrelated to his flight.

Q *What was the first animal to ride a rocket?*

A A rhesus monkey named Albert, launched in a USAF V-2 rocket in 1948. He also became the first animal to die in a rocket.

Q *Name the chimp who flew in a Mercury craft before Alan Shepard did.*

A Animal Subject 65. His handlers called the chimp Chang, but NASA vetoed the use of that name for fear of angering "every Chinese laundryman in America, not to mention the Taiwanese" (as if they wouldn't be offended by being called laundrymen). An acronym for the Holloman Aeromedical Research Laboratory gave the chimp the cute, acceptable, and American name of Ham. Officials later realized the headlines that could be made of "Ham Bakes in Mercury."

Q *Who were Anita and Arabella?*

A A pair of spiders put aboard Skylab 3 to see if weightlessness affects web-spinning. It didn't.

Q *Who were the first "animalnauts" to survive re-entry, landing, and recovery?*

A Belka ("Squirrel") and Strelka ("Arrow"), the two puppies aboard Sputnik 5, August 19, 1960. One of Belka's puppies was later given to Caroline Kennedy.

Q *Name the ship that recovered the chimpanzee Ham after his spaceflight.*

A The USS *Donner,* the prime recovery ship. It is seen here moving to its station in the Atlantic prior to the flight.

> **No Monkey Business Here!** The chimpanzees that NASA trained at Holloman Air Force Base became experts in learning set operations. In order to receive a banana-flavored pellet they had to discharge a lever exactly 50 times. In all cases they were able to quickly pull the lever 45 times and then carefully pull the final 5 and cup their hands to catch the pellet. One chimp actually attained an accuracy rate of 99.6%, making only 28 mistakes in 7,000 pulls of the lever.

Q *A chimp named Ham rode in Mercury-Redstone-2 on January 31, 1961. Name the chimp that rode in Mercury-Atlas-5 on November 29, 1961.*

A Enos, the brightest of the lot, whose name means "man" in Greek. His flight lasted for 3 hours, 20 minutes, and 59 seconds, considerably longer than the 16-minute, 39-second flight of Ham.

Q *What was the first living thing ever strapped into a Mercury capsule for a test?*

A A pig. The body tissue in pigs resembles the tissue in humans. The pig in the capsule was dropped down a shaft to determine what problems might be expected in the event of a "hard" landing on the ground rather than a "soft" landing in the sea. The pig survived nicely.

Q *Name the rhesus and squirrel monkeys that were launched on top of a Jupiter rocket in 1959.*

A Able and Baker. Baker outlived her spacemate by more than twenty years and was a resident of the Alabama Space and Rocket Center until she died in 1984.

Q *What did the message "Yes, Yes" signify?*

A It was an answer to the question "Are the monkeys safe?" after the 1959 launch of two squirrel monkeys atop a Jupiter rocket. Radio relays between the Cape and the recovery ships were bad, and the brief message was correctly interpreted as meaning that both monkeys had survived.

Q *Name the other four chimpanzees that arrived at Cape Canaveral with Enos as candidates for the Mercury-Atlas flight 5, which was launched on November 29, 1961.*

A Duane, Rocky, Jim, and the only one to have flown previously, Ham.

Q *In how many languages did the Voyager record carry greetings?*

A Fifty-four human languages, from Sumerian to Swahili, and one whale language.

Q *Name the first manmade object to leave the solar system.*

A Pioneer 10, shown here, which passed the orbits of all nine known planets on June 13, 1983. Launched March 1972, Pioneer 10 was the first craft to navigate the treacherous asteroid belt, and in 1973 it transmitted the first close photos of Jupiter. It will remain in contact until 1991.

Q *What is the significance of this picture?*

A It shows the first message sent out of the solar system, on 6-by-9-inch plaques attached to Pioneers 10 and 11, launched in March 1972 and April 1973. The message, designed by Drs. Frank Drake and Carl Sagan, both of Cornell, describes the spacecraft, humans, our solar system, and the hydrogen atom. The "spider" at left is a cosmic roadmap, showing our location in relation to known pulsars. Any civilization that deciphers it will be extremely intelligent, since most of the professors at Cornell couldn't figure it out.

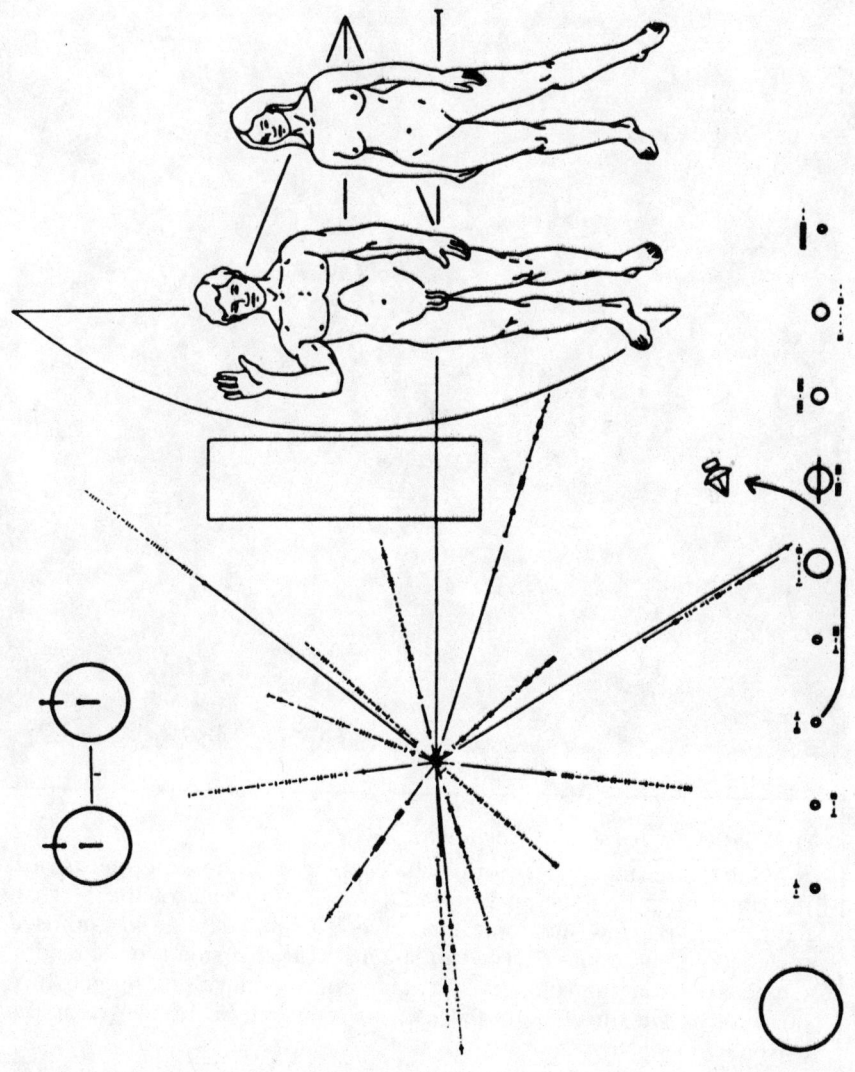

Q *What is the Barsoom Inscription?*

A Viking 1's camera picked up what appeared to be an upper-case *B* inscribed on a boulder on Chryse. It was just a trick of the Martian light, but for a few moments there were some excited and confused scientists in Pasadena. Carl Sagan suggested that it stood for Barsoom, which was the native name for Mars in the novels of Edgar Rice Burroughs. The mystery initial can be seen below the arrow at the extreme left.

Q *Name the monkey that was killed when the Mercury spacecraft he was riding in was destroyed when it went off course after launch on November 10, 1961.*

A Goliath, a 1½-pound squirrel monkey. The range safety officer in Cape Canaveral pressed the destruct button to terminate the flight when it became obvious that the Atlas rocket was out of control.

Q *Name the first animal to orbit the Earth.*

A A part-Samoyed puppy generally known as Laika ("Barker"), also known as Kudryavka ("Curly") or Limonchik ("Little Lemon"), launched November 3, 1957, aboard Sputnik 2. According to some sources, the Russians poisoned her by remote control after ten days rather than let her suffocate.

Q *What caused the communications blackout during the chimpanzee Enos's flight in MA-5?*

A The telephone cable line between Cape Canaveral and Point Arguello was snapped by a farm tractor in Arizona. For a brief period contact between the coasts was disrupted, but it was quickly restored via another cable.

Q *Prior to the U.S. launching of a Saturn 5 rocket, what was the heaviest payload ever lofted by man?*

A Thirteen tons, by the Soviets. The first Saturn 5, with an Apollo spacecraft, weighed 140 tons.

Q *What is recognized as the first scientific test of a liquid-fueled rocket?*

A The March 16, 1926, experiment by Robert H. Goddard. His 10-foot-long rocket ascended 41 feet and reached a speed of 60 mph. The fuel was gasoline and liquid oxygen, ignited with a blowtorch. The location was a farm owned by Goddard's aunt, near Auburn, Massachusetts.

Q *When and where did the first man-made object, other than a bullet, achieve supersonic speed?*

A On October 3, 1942, German scientists at Peenemünde launched a 13½-ton, 47-foot V-2 rocket that reached a speed of 3,500 mph and an altitude of over 50 miles. Wernher von Braun, who later contributed greatly to the U.S. space program, was part of the German team.

Q *Have the Russians ever retrieved lunar samples?*

A Yes. Although not as world-shaking as a manned lunar landing, the Soviet Luna-16 probe brought Moon rocks back to Earth by robot in September 1970.

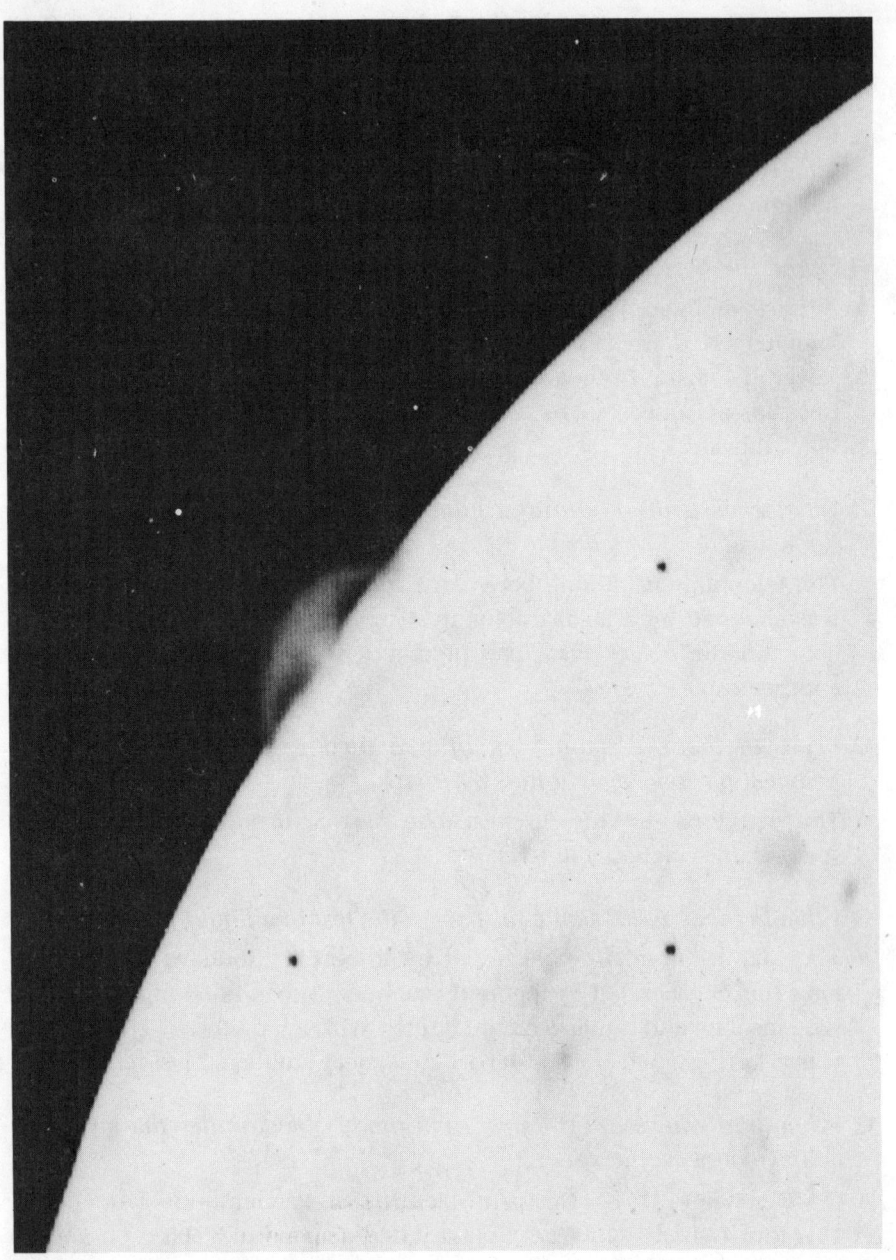

Q *How many places in the solar system have active volcanoes?*

A Three: Earth, Venus, and Io, one of the Jovian moons. Volcanoes discovered on Mars appear to be dormant or extinct. Shown here is a volcanic eruption on Io. The lack of atmosphere causes the ejecta to fall in a regular umbrella-shaped pattern, unlike the situation on Earth.

Small Is Better Because of miniaturization it was possible to contain nearly 9 million separate parts in the Apollo-Saturn space vehicles. In addition, the Saturn rocket contained 2,250 tons of fuel ... which it consumed in 2.5 minutes.

Q *When was the first satellite launched?*

A Sputnik 1, a 22-inch, 184-pound instrument package, was launched on October 4, 1957. Circling the Earth every 1 hour and 36 minutes at an altitude of 560 miles, it provided information on ionospheric temperature and density. It shook U.S. complacency about its technological supremacy, and got America into the space business.

Q *Name the first successful Mars probe.*

A Mariner 4, launched November 28, 1964. Six months later, it beamed twenty photographs back to Earth.

Q *What was the first artificial satellite to orbit the Sun?*

A The Soviet Luna 1, launched January 2, 1958. Intended to strike the Moon, it missed and went into orbit about the Sun by default. The Soviets immediately renamed it *Mechta,* or "Daydream," and pronounced it a great success.

Q *What minor error led to the failure of an American space probe?*

A In 1962, a computer programmer omitted a minus sign from coded instructions to the Venus-bound Mariner 1. The omission caused Mariner to go off course. It was destroyed by remote control.

Q *Name the first spacecraft to photograph the far side of the Moon.*
A Luna 3, October 4, 1959.

Q *What was the first spacecraft to go beyond Earth orbit?*
A Luna 1, called *Mechta,* the first Soviet attempt to hit the Moon.

Q *When was the first soft landing on the Moon?*
A January 31, 1966, when the Soviet Luna 9 spacecraft settled down to take the first panoramic photos of another body.

Q *Name the world's first commercial communications satellite.*
A *Early Bird,* launched on April 6, 1965. Commercial service began on June 28.

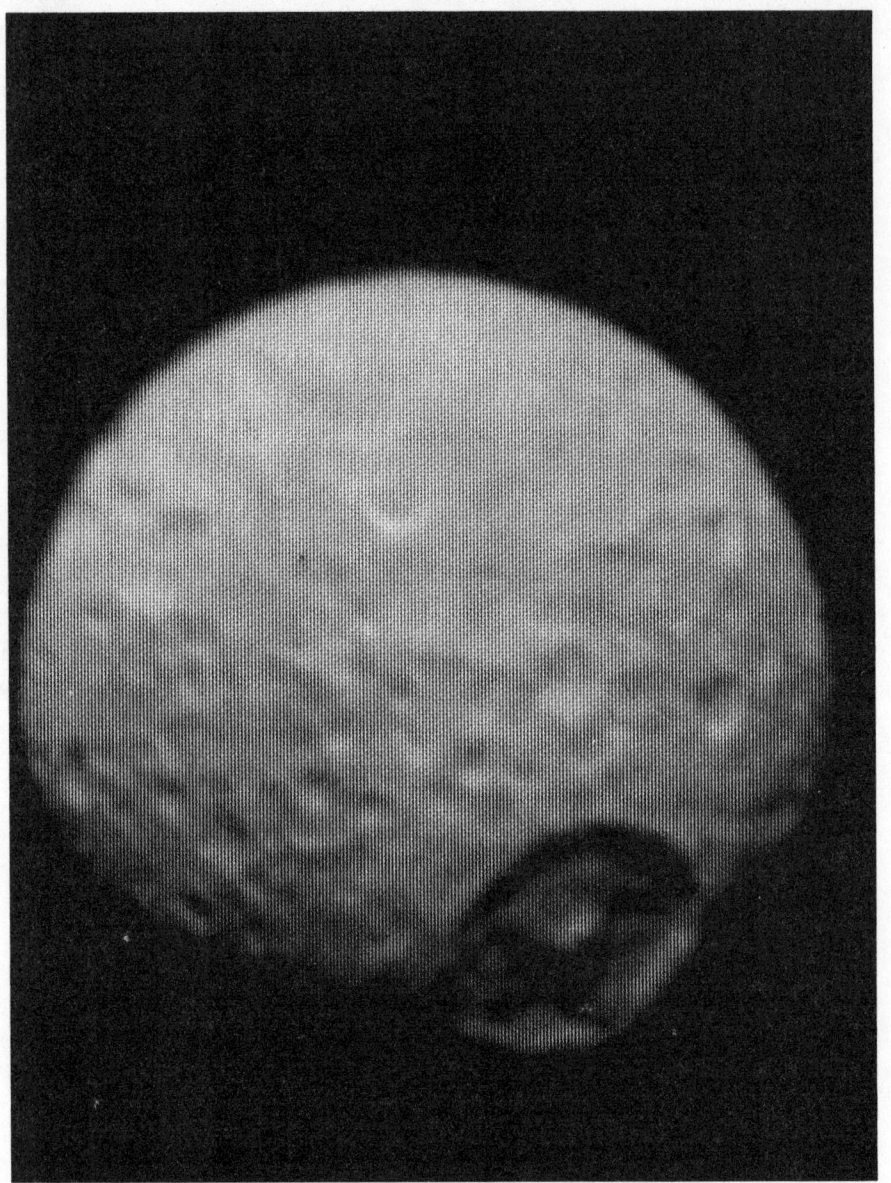

Q *What is a "Death Star"?*

A Named for their resemblance to the ultimate weapons in the *Star Wars* movies (which were recalled when they kept exploding), Death Stars are planets that have received a major meteoroid impact, carving out a large crater on one side. Mimas (shown here) and Tethys, two moons of Saturn, and the planet Mercury are among the Death Stars already discovered.

The Scoop That Wasn't *New York Times* science correspondent Walter Sullivan was convinced that a Soviet space mission was just around the corner. He filed a story predicting the spaceflight on Friday, October 4, 1957. A few hours later the Soviets launched Sputnik. His story didn't run.

Q *Which spacecraft took the first detailed photos of the planet Mercury?*

A Mariner 10, launched November 1, 1973. Mariner first did a flyby of Venus, then went into an orbit around the Sun calculated to bring it close to Mercury. The first photos were transmitted on March 29, 1972, later ones on September 21, 1974, and March 16, 1975.

Q *Where is the Caloris Basin?*

A On Mercury. A circular basin 8,125 miles in diameter and ringed by half-mile-high mountains, it was probably formed by collision with a large meteorite.

Q *Name the first spacecraft to land on Venus.*

A Venera 5. Earlier Soviet attempts to reach Venus had either failed in transit or imploded in the hot, high-pressure atmosphere. Veneras 5 through 10 sent back information for up to 50 minutes before frying.

Q *In what year were the first TV pictures transmitted from the surface of another planet?*

A In December 1975, when the Soviet Venera 10 probe landed on Venus.

Q *Does it rain on Venus?*

A Yes, it rains sulphuric acid in an atmosphere of mostly carbon dioxide. But it never hits ground; the temperature is so hot that it evaporates on the way down and returns to the clouds.

Q *When did the first photographic flyby of Mars take place?*

A On July 14, 1965, when the Mariner 4 spacecraft came within 6,250 miles of Mars. It sent back pictures of a cratered, desert world, quenching hopes for discovery of a technological civilization on the red planet.

Q *What was the political significance of the Viking 1 landing date?*

A It was originally slated to land on July 4, 1976, for America's bicentennial. Orbital photos showed the landing site to be rockier than anticipated, so landing was delayed 2½ weeks until another site was found.

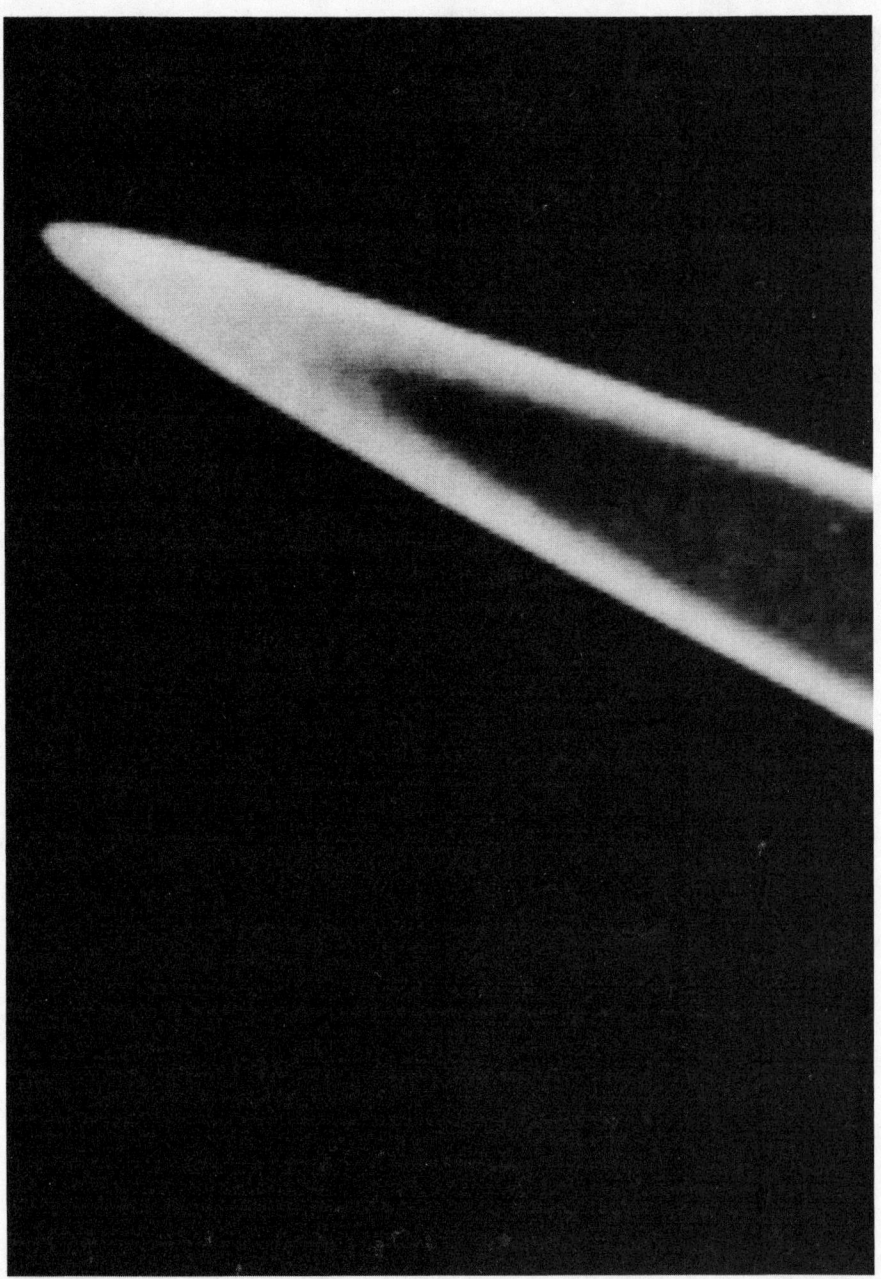

Q *Which planets are surrounded by rings?*

A Uranus, Saturn, and Jupiter. The ring around Jupiter, seen here, was discovered accidentally by the Voyager 1 probe. It's believed all large planets have rings.

> **Is Anybody There?** When the world's first communications satellite was placed in orbit on December 18, 1959, great care was taken to make sure that the first message would be something historic and appropriate, in the form of a recorded statement from President Dwight D. Eisenhower. However, the first voice transmission actually picked up was that of a technician at the tracking facility in Goldstone, California, calling to a colleague in Holmdel, New Jersey: "Hey Walt! Can you hear me?"

Q *Name the first mission to map the entire surface of Mars.*

A Mariner 9, in 1971. The mapping was necessary for planning the Viking mission.

Q *Who first described the "canals" of Mars?*

A Italian astronomer Giovanni Schiaparelli, in 1877. He called the dark streaks in the red surface *canali,* or channels. The English translation to "canals" suggested a structure built by intelligent creatures.

Q *What were the "oases" of Mars?*

A Dark spots where the "canals" crossed, reported by American astronomer William Pickering.

Q *What causes the seasonal streaking and color changes on Mars?*

A Wind. Prior to the Mariner series, it had been thought the changes were advancing vegetation. Carl Sagan of Cornell University predicted that the changes were really caused by seasonal winds blowing lighter sand away from darker rock underneath. Later missions bore this theory out.

Q *Why was it decided to land the Viking spacecraft in low-elevation sites?*

A The Martian atmosphere is so thin that even with a large parachute, a lot of distance is needed to slow down the lander for a safe touchdown.

Q *Name the landing sites for Viking 1 and 2.*

A Chryse and Utopia. Other areas were more likely to harbor life, but landings there would have been riskier.

Q *What did Project Viking determine about life on Mars?*

A The official report concludes, "There is no evidence of life on Mars." This is controversial since three out of four experiments gave positive results.

Q *When was the first soft landing on Mars?*
A July 20, 1976, when the Viking 1 probe landed. The first picture it transmitted was of its own footpad, so researchers could tell if it was about to sink into the Martian soil. It didn't.

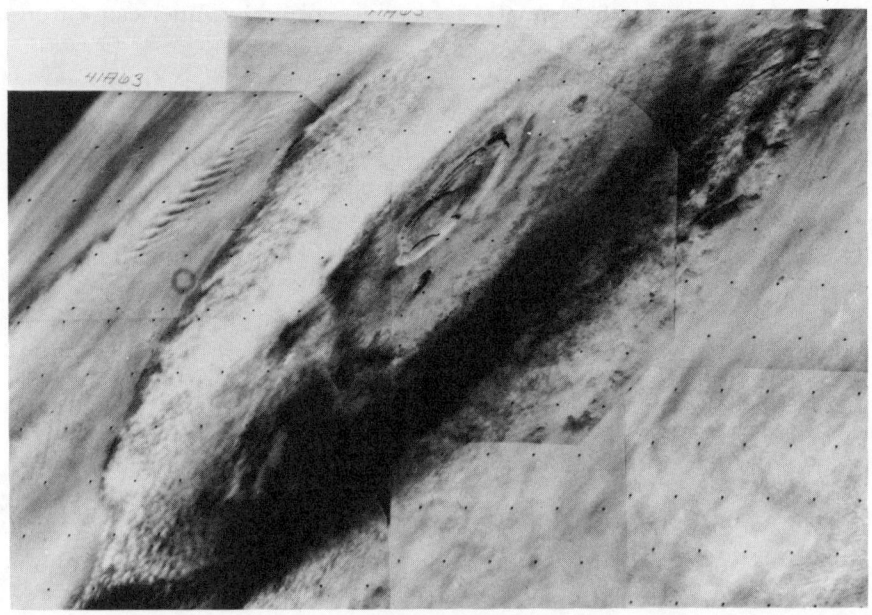

Q *Where is the largest volcano in the solar system?*
A On Mars. Called Olympus Mons, it is three times higher than Everest and covers an area the size of Arizona. It's also the largest *inactive* volcano in the solar system.

Q *Who made the first commercial phone call carried by satellite?*

A President Lyndon Johnson, to Prime Minister Harold Wilson of Great Britain, on June 28, 1965.

Q *On which mission was the first picture of the entire Earth taken?*

A Lunar Orbiter 1 produced a photo of the "gibbous" Earth and a large portion of the Moon's arc on August 14, 1966.

Q *Who popularized the idea of Mars as a dying planet with a civilization in search of water?*

A French writer Camille Flammarion in *Planet Mars*, in 1892, and astronomer Percival Lowell with *Mars* in 1894. Lowell suggested that the canals were an engineering scheme to bring water from the polar ice cap to the parched equator, and that the Martians were older and more advanced than we mere humans. The theme of aliens from a drought-stricken world scheming to seize or escaping to Earth has been repeated many times: see *The Man Who Fell to Earth*, *The Invaders*, and the recent series *V*.

Q *What supernatural assistance was given to the Navy's Vanguard satellite?*

A After the first launch attempt ended in a nationally televised explosion, technicians welded several St. Christopher medals to the guidance section and launched successfully on St. Patrick's Day, 1958. Too late to make the record books, though. The Army's Redstone rocket had lifted Explorer 1 into orbit two months earlier.

Q *Which experiment led to the conclusion that there was no life on Mars?*

A The search for complex carbon-based molecules, which are essential for life as we know it. Without the building blocks of life, scientists had to consider results from the other experiments invalid.

Q *What was the last Playboy Redstone?*

A During the launch of the fourth Redstone rocket from Cape Canaveral, the crews had maintained the military tradition of decorating the bird with cheesecake artwork, done on posterboard and taped to the main stage. On ascent, the wind ripped it off, as expected. But when the tracking radar locked onto the falling poster and ignored the soaring rocket, Dr. von Braun ordered an end to the practice.

Appendixes

THE ASTRONAUT CORPS

The listing that follows includes brief biographies of all 127 people selected as astronaut candidates by NASA in Groups 1 through 9 between 1959 and 1980. The information appears in the following order: name, date of birth, hometown, military status, earned degrees, field, astronaut group and date, flights made, personal data. Not all data applies to each person. The ranks given for astronauts not on flight status are their ranks when they left NASA.

Aldrin, Buzz (Jan. 20, 1930, Montclair, N.J.)—Colonel, USAF (Ret.); B.S. (mechanical engineering), M.D.; Group 3, October 1963; Gemini 12, Apollo 11; resigned from NASA, July 1971; retired from Air Force active duty, March 1, 1972.

Allen, Joseph P. (June 27, 1937, Crawfordsville, Ind.)—Civilian; B.A. (mathematics and physics), M.S. and Ph.D. (physics); flight, mission specialist; Group 6, August 1967; NASA Assistant Administrator for Legislative Affairs, August 1975–August 1978; served as a mission specialist on STS-5, the first operational flight of the space shuttle orbiter *Columbia*, Nov. 11–16, 1982.

Anders, William A. (Oct. 17, 1933, Hong Kong)—Colonel, USAF (Reserve) (now Major General); B.S. from U.S. Naval Academy, M.S. (nuclear engineering); Group 3, October 1963; Apollo 8; resigned from NASA, September 1969. Vice President, Textron Inc.

Armstrong, Neil A. (Aug. 5, 1930, Wapakoneta, Ohio)—Civilian; B.S. and M.S. (aeronautical engineering); Group 2, September 1962; Gemini 8, Apollo 11; was Deputy Associate Administrator, Aeronautics, NASA Headquarters Office of Advanced Research and Technology, 1970–1971; resigned from NASA, 1971. Chairman, Computing Technologies for Aviation, Charlottesville, Va.

Bagian, James P. (Feb. 22, 1952, Philadelphia, Pa.)—Civilian; B.S. (mechanical engineering), M.D.; flight, mission specialist; Group 9, August 1981.

Bassett, Charles A. (Dec. 30, 1931, Dayton, Ohio)—Major, USAF; B.S. (electrical engineering); deceased; Group 3, October 1963; died in T-38 jet crash with Elliott See, Feb. 28, 1966, Lambert Municipal Airport, St. Louis, Mo.

Bean, Alan L. (March 15, 1932, Wheeler, Texas, but considers Fort Worth, Texas, his hometown)—Captain, USN (Ret.); B.S. (aeronautical engineering); Group 3, October 1963; Apollo 12, Skylab 3; retired from the Navy, October 1975; resigned from NASA, June 1981, to devote his full time to painting.

Blaha, John E. (Aug. 26, 1942, San Antonio, Texas)—Colonel, USAF; B.S. (engineering science), M.S. (astronautical engineering); flight, pilot; Group 9, August 1981.

Bluford, Guion S., Jr. (Nov. 22, 1942, Philadelphia, Pa.)—Lt. Colonel, USAF; B.S. (aerospace engineering), M.S. (aerospace engineering); flight, mission specialist; Group 8, August 1979; as mission specialist for STS-8, Sept. 30, 1983.

Bobko, Karol J. (Dec. 23, 1937, New York City)—Colonel, USAF; B.S. from U.S. Air Force Academy, M.S. (aerospace engineering); flight, pilot; Group 7, August 1969; pilot on STS-6, April 4–9, 1983.

Bolden, Charles F., Jr. (Aug. 19, 1946, Columbia, S.C.)—Major USMC; B.S. (electrical science), M.S. (systems management); flight, pilot; Group 9, August 1981.

Borman, Frank (March 14, 1928, Gary, Ind.)—Colonel, USAF (Ret.); B.S. from U.S. Military Academy, M.S. (aeronautical engineering); Group 2, September 1962; Gemini 7, Apollo 8; retired from Air Force and resigned from NASA, July 1970. Chairman, President, and Chief Executive Officer, Eastern Airlines, Miami, Fla.

Brand, Vance D. (May 9, 1931, Longmont, Colo., but considers Gainesville, Ga., to be his hometown)—Civilian; B.S. (business and aeronautical engineering), M.S. (business administration); flight, pilot; Group 5, April 1966; Apollo-Soyuz Test Project; commander of STS-5, the fifth flight of space shuttle *Columbia*, Nov. 11–16, 1982; commander on STS-11 (41-B), Feb. 3–10, 1984.

Brandenstein, Daniel C. (Jan. 17, 1943, Watertown, Wis.)—Commander, USN; B.S. (mathematics and physics); flight, pilot; Group 8, August 1979; pilot for STS-8, Sept. 30, 1983.

Bridges, Roy D., Jr. (July 19, 1943, Atlanta, Ga., but considers Gainesville, Ga., his hometown)—Lt. Colonel, USAF; B.S. (engineering science), M.S. (astronautics); flight, pilot; Group 9, August 1981.

Buchli, James F. (June 20, 1945, New Rockford, N.D., but considers Fargo, N.D., his hometown)—Major, USMC; B.S. (aeronautical engineering), M.S. (aeronautical engineering systems); flight, mission specialist; Group 8, August 1979.

Bull, John S. (Sept. 25, 1934, Memphis, Tenn.)—Lt. Commander, USN (Ret.); B.S. (mechanical engineering), M.S. and Ph.D. (aeronautical engineering); Group 5, April 1966; resigned from NASA, July 1968; withdrew from astronaut program and the Navy because of pulmonary disease. Member, Guidance and Navigation Branch, Ames Research Center, Mountain View, Calif.

Carpenter, M. Scott (May 1, 1925, Boulder, Colo.)—Commander, USN (Ret.); B.S. (aeronautical engineering); Group 1, April 1959; Mercury 7; joined U.S. Navy SEALAB program in 1967; resigned from NASA, August 1967; retired from Navy, July 1969.

Carr, Gerald P. (Aug. 22, 1932, Denver, Colo., but considers Santa Ana, Calif., his hometown)—Colonel, USMC (Ret.); B.S. (mechanical engineering), B.S. and M.S. (aeronautical engineering); Group 5, April 1966; Skylab 4; resigned from NASA, June 1977; retired from Marine Corps, September 1975. Senior Consultant, Applied Research, Inc., Houston.

Cernan, Eugene A. (March 14, 1934, Chicago, Ill.)—Captain, USN (Ret.); B.S. (electrical engineering), M.S. (aeronautical engineering); Group 3, October 1963; Gemini 9, Apollo 10, Apollo 17; resigned from NASA and retired from Navy, July 1976. Cernan Energy Corp., Houston.

Chaffee, Roger B. (Feb. 15, 1935, Grand Rapids, Mich.)—Lt. Commander, USN; B.S. (aeronautical engineering); deceased; Group 3, October 1963; died in Apollo spacecraft fire, Kennedy Space Center, Jan. 27, 1967.

Chang, Franklin R. (April 5, 1950, San Jose, Costa Rica)—Civilian; B.S. (mechanical engineering), Ph.D. (applied plasma physics); flight, mission specialist; Group 9, August 1981.

Chapman, Philip K. (March 5, 1935, Melbourne, Australia)—Civilian; B.S. (physics and mathematics), M.S. (aeronautics and astronautics), D.Sc. (instrumentation); Group 6, August 1967; resigned from NASA, July 1972. Arthur D. Little, Inc., Cambrige, Mass.

Cleave, Mary L. (Feb. 5, 1947, Southampton, N.Y.)—Civilian; B.S. (biological sciences), M.S. (microbial ecology), Ph.D. (civil and environmental engineering); flight, mission specialist; Group 9, August 1981.

Coats, Michael L. (Jan. 16, 1946, Sacramento, Calif., but considers Riverside, Calif., his hometown)—Commander, USN; B.S. from U.S. Naval Academy, M.S. (admin. of science and technology), M.S. (aeronautical engineering); flight, pilot; Group 8, Aug. 1979. Pilot, 41-D, Aug. 30–Sept. 6, 1984.

Collins, Michael (Oct. 31, 1930, Rome, Italy)—Colonel (now Major General, USAFR Ret.); B.S. from U.S. Military Academy; Group 3, October 1963; Gemini 10, Apollo 11; resigned from NASA, January 1970. Michael Collins Assocs., Washington, D.C.

Conrad, Charles, Jr. (June 2, 1930, Philadelphia, Pa.)—Captain, USN (Ret.); B.S. (aeronautical engineering); Group 2, September 1962; Gemini 5, Gemini 11, Apollo 12, Skylab 2; resigned from NASA and retired from Navy, December 1973. Senior Vice President, Marketing, Douglas Aircraft Co., Long Beach, Calif.

Cooper, L. Gordon (March 6, 1927, Shawnee, Okla.)—Colonel, USAF (Ret.); B.S. (aeronautical engineering), D.Sc.; Group 1, April 1959; Mercury 9, Gemini 5; retired from NASA and Air Force, July 1970.

Covey, Richard O. (Aug. 1, 1946, Fayetteville, Ark., but considers Fort Walton Beach, Fla., his hometown)—Lt. Colonel, USAF; B.S. (engineering sciences), M.S. (aeronautics and astronautics); flight, pilot; Group 8, August 1979.

Creighton, John O. (April 28, 1943, Orange, Texas, but considers Seattle, Wash., his hometown)—Commander, USN; B.S. from U.S. Naval Academy, M.S. (administration of science and technology); flight, pilot; Group 8, August 1979.

Crippen, Robert L. (Sept. 11, 1937, Beaumont, Texas)—Captain, USN; B.S. (aerospace engineering); flight, pilot; Group 7, August 1969; pilot on STS-1, April 12–14, 1981; commander on STS-7, June 18–24, 1983; commander for 41-C, April 4–12, 1984; commander 41-G, Oct. 5–13, 1984.

Cunningham, Walter (March 16, 1932, Creston, Iowa)—Civilian; B.A., M.A. and Ph.D. (physics); Group 3, October 1963; Apollo 7; resigned from NASA, August 1971. The Capital Group, Houston.

Duke, Charles M., Jr. (Oct. 3, 1935, Charlotte, N.C.)—Brig. General, USAF (Reserve); B.S. (naval sciences), M.S. (aeronautics); Group 5, April 1966; Apollo 16; resigned from NASA, December 1975; resigned from Air Force, January 1976. Duke Investments, and President, Southwest Wilderness Art, Inc.

Dunbar, Bonnie J. (March 3, 1949, Sunnyside, Wash.)—Civilian; B.S. and M.S. (ceramic engineering), presently doctoral candidate in biomedical engineering; flight, mission specialist; Group 9, August 1981.

Eisele, Donn F. (June 23, 1930, Columbus, Ohio)—Colonel, USAF (Ret.); M.S. (astronautics); Group 3, October 1963; Apollo 7; resigned from NASA and retired from Air Force, July 1972. Was technical assistant for manned space flight, NASA Langley Research Center, Hampton, Va., 1970–1972. Oppenheimer and Co., Inc., Ft. Lauderdale, Fla.

England, Anthony W. (May 15, 1942, Indianapolis, Ind., but considers Fargo, N.D., his hometown)—Civilian; B.S. and M.S. (geology and physics), Ph.D. (earth and planetary sciences); flight, mission specialist; Group 6, August 1967; resigned in August 1972 to accept position with the U.S. Geological Survey; rejoined NASA in 1979 as a scientist-astronaut; designated as a mission specialist on STS-24 (Spacelab 2).

Engle, Joe H. (Aug. 26, 1932, Chapman, Kans.)—Colonel, USAF; B.S. (aeronautical engineering); flight, pilot; Group 5, April 1966; commanded *Enterprise* space shuttle free-flight approach and landing tests 2 and 4, Sept. 13 and Oct. 12, 1977; commander STS-2, Nov. 12–14, 1981; Deputy Associate Administrator for Space Flight at NASA Headquarters, April–December 1982; returned to astronaut duties at the Johnson Space Center to begin training for his next shuttle flight.

Evans, Ronald E. (Nov. 10, 1935, St. Francis, Kans.)—Captain, USN (Ret.); B.S. (electrical engineering), M.S. (aeonautical engineering); Group 5, April 1966; Apollo 17; retired from Navy, April 1976; resigned from NASA, March 1977. Director, Space Systems Marketing for Sperry Flight Systems, Phoenix, Ariz.

Fabian, John M. (Jan. 28, 1939, Goosecreek, Texas, but considers Pullman, Wash., his hometown)—Colonel, USAF; B.S. (mechanical engineering), M.S. (aerospace engineering), Ph.D. (aeronautics and astronautics); flight, mission specialist; Group 8, August 1979; mission specialist on STS-7, June 18–24, 1983.

Fisher, Anna L. (Aug. 24, 1949, St. Albans, N.Y., but considers San Pedro, Calif., her hometown)—Civilian; B.S. (chemistry), M.D.; flight, mission specialist; Group 8, August 1979. Mission specialist, 51-A, Nov. 8–16, 1984.

Fisher, William F. (April 1, 1946, Dallas, Texas)—Civilian; B.A., M.S. (engineering), M.D.; flight, mission specialist; Group 9, August 1981.

Freeman, Theodore C. (Feb. 18, 1930, Haverford, Pa.)—Captain, USAF; B.S. from U.S. Naval Academy, M.S. (aeronautical engineering); deceased; Group 3, October 1963; died in T-38 crash, Ellington AFB, Houston, Oct. 31, 1964.

Fullerton, C. Gordon (Oct. 11, 1936, Rochester, N.Y.)—Colonel, USAF; B.S. and M.S. (mechanical engineering); flight, pilot; Group 7, August 1969; piloted *Enterprise* space shuttle free-flight approach and landing tests 1, 3, and 5 on Aug. 12, Sept. 23, and Oct. 26, 1977; pilot for STS-3, March 22–30, 1982.

Gardner, Dale A. (Nov. 8, 1948, Fairmont, Minn., but considers Clinton, Iowa, his hometown)—Lt. Commander, USN; B.S. (engineering physics); flight, mission specialist; Group 8, August 1979; mission specialist, STS-8, Sept. 30–Oct. 5, 1983.

Gardner, Guy S. (Jan. 6, 1948, Alta Vista, Va.)—Major, USAF; B.S. (engineering sciences), M.S. (astronautics); flight, pilot; Group 9, August 1981.

Garriott, Owen K. (Nov. 22, 1930, Enid, Okla.)—Civilian; B.S., M.S., and Ph.D. (electrical engineering); flight, mission specialist; Group 4, June 1965; Skylab 3; mission specialist for STS-9 (Spacelab 1), Nov. 28–Dec. 8, 1983.

Gibson, Edward G. (Nov. 8, 1936, Buffalo, N.Y.)—Civilian; B.S., M.S. (engineering), Ph.D. (engineering and physics); Group 4, June 1965; Skylab 4; resigned, November 1974, then rejoined NASA in March 1977, and resigned again in October 1980. Advanced Systems Manager, TRW Inc., Redondo Beach, Calif.

Gibson, Robert L. (Oct. 30, 1946, Cooperstown, N.Y., but considers Lakewood, Calif., his hometown)—Lt. Commander, USN; B.S. (aeronautical engineering); flight, pilot; Group 8, August 1979; pilot, STS-11, (41-B), Feb. 3–10, 1984.

Givens, Edward G. (Jan. 5, 1930, Quanah, Texas)—Major, USAF; B.S. (naval sciences); deceased; Group 5, April 1966; died in an automobile accident near Houston, June 6, 1967.

Glenn, John H., Jr., (July 18, 1921, Cambridge, Ohio)—Colonel, USMC (Ret.); B.S. (engineering); Group 1, April 1959; Mercury 6; resigned from NASA, January 1964. Elected to the U.S. Senate in November 1974, where he now serves.

Gordon, Richard F., Jr. (Oct. 5, 1929, Seattle, Wash.)—Captain, USN (Ret.); B.S. (chemistry); Group 3, October 1963; Gemini 11, Apollo 12; retired from Navy and resigned from NASA, January 1972. President, Astro Systems & Engineering, Inc., Los Angeles, Calif.

Grabe, Ronald J. (June 13, 1945, New York City)—Major, USAF; B.S. (engineering science); flight, pilot; Group 9, August 1981.

Graveline, Duane E. (March 2, 1931, Newport, Vt.)—Civilian; M.D.; resigned, August 1965; Group 4, June 1965.

Gregory, Frederick D. (Jan. 7, 1941, Washington, D.C.)—Lt. Colonel, USAF; B.S. from U.S. Air Force Academy, M.S. (information systems); flight, pilot; Group 8, August 1979; designated to serve as pilot for STS-18 (Spacelab 3).

Griggs, S. David (Sept. 7, 1939, Portland, Ore.)—Civilian; B.S. from U.S. Naval Academy, M.S. (administration); flight, pilot; Group 8, August 1979.

Grissom, Virgil I. (April 3, 1926, Mitchell, Ind.)—Lt. Colonel, USAF; B.S. (mechanical engineering); deceased; Group 1, April 1959; Mercury 4, Gemini 3; died in Apollo spacecraft fire at Kennedy Space Center, Jan. 27, 1967.

Haise, Fred W., Jr. (Nov. 14, 1933, Biloxi, Miss.)—Civilian; B.S. (aeronautical engineering); Group 5, April 1966; Apollo 13; commanded *Enterprise* space shuttle free-flight approach and landing tests 1, 3, and 5, Aug. 12, Sept. 23, and Oct. 26, 1977; resigned from NASA, June 1979. Vice President, Space Programs, Grumman Aerospace Corp., Titusville, Fla.

Hart, Terry J. (Oct. 27, 1946, Pittsburgh, Pa.)—Civilian; B.S. and M.S. (mechanical engineering), M.S. (electrical engineering); flight, mission specialist; Group 8, August 1979; selected to serve as a mission specialist on STS-13.

Hartsfield, Henry W., Jr. (Nov. 21, 1933, Birmingham, Ala.)—Colonel, USAF (Ret.); B.S. (physics), M.S. (engineering science); flight, pilot; Group 7, August 1969; retired from Air Force, August 1977. Pilot for STS-4, June 27–July 4, 1982; commander, STS-12, (41-D), Aug. 30–Sept. 5, 1984.

Hauck, Frederick H. (April 11, 1941, Long Beach, Calif., but considers Winchester, Mass., and Washington, D.C., his hometowns)—Captain, USN; B.S. (physics), M.S. (nuclear engineering); flight, pilot; Group 8, August 1979; pilot for STS-7, June 18–24, 1983; commander 51-A, Nov. 8–16, 1984.

Hawley, Steven A. (Dec. 12, 1951, Ottawa, Kans., but considers Salina, Kans., his hometown)—Civilian; B.A. (physics and astronomy), Ph.D. (astronomy and astrophysics); flight, mission specialist; Group 8, August 1979; mission specialist on STS-12 (41-D), Aug. 30–Sept. 5, 1984.

Henize, Karl G. (Oct. 17, 1926, Cincinnati, Ohio)—Civilian; B.A. (mathematics), M.A. and Ph.D. (astronomy); flight, mission specialist; Group 6, August 1967; selected to serve as a mission specialist on STS-24 (Spacelab 2).

Hilmers, David C. (Jan. 28, 1950, Clinton, Iowa, but considers DeWitt, Iowa, his hometown)—Captain, USMC; B.S. (mathematics), M.S. (electrical engineering); flight, mission specialist; Group 9, August 1981.

Hoffman, Jeffrey A. (Nov. 2, 1944, Brooklyn, N.Y., but considers Scarsdale, N.Y., his hometown)—Civilian; B.A. (astronomy), Ph.D. (astrophysics); flight, mission specialist; Group 8, August 1979.

Holmquest, Donald L. (April 7, 1939, Dallas, Texas)—Civilian; M.D., B.S. (electrical engineering), Ph.D. (physiology); Group 6, August 1967; took leave of absence May 1971 to hold position of Assistant Professor of Radiology and Physiology, Baylor College of Medicine, Houston; resigned from NASA, September 1973. Now practices medicine on a full-time basis.

Irwin, James B. (March 17, 1930, Pittsburgh, Pa.)—Colonel, USAF (Ret.); B.S. (naval science), M.S. (aeronautical and instrumentation engineering); Group 5, April 1966; Apollo 15; resigned from NASA, August 1972. Chairman of Board, High Flight Foundation, Colorado Springs, Colo.

Kerwin, Joseph P. (Feb. 19, 1932, Oak Park, Ill.)—Captain, MC, USN; B.A. (philosophy), M.D.; flight, mission specialist; Group 4, June 1965; Skylab 2; currently NASA representative in Australia. At conclusion of this two-year assignment, Kerwin will return to astronaut duties at the Johnson Space Center.

Leestma, David C. (May 6, 1949, Muskegon, Mich., but considers Tustin, Calif., his hometown)—Lt. Commander, USN; B.S. and M.S. (aeronautical engineering); flight, mission specialist; Group 9, August 1981. Mission specialist, 41-G, Oct. 5–13, 1984.

Lenoir, William B. (Mar. 14, 1939, Miami, Fla.)—Civilian; B.S., M.S., and Ph.D. (electrical engineering); flight, mission specialist; Group 6, August 1967; mission specialist on STS-5, Nov. 11–16, 1982.

Lind, Don L. (May 18, 1930, Midvale, Utah)—Civilian; B.S. and Ph.D. (physics); flight, pilot; Group 5, April 1966; designated as a mission specialist on STS-18 (Spacelab 3).

Llewellyn, John A. (April 22, 1933, Cardiff, Wales)—Civilian; Ph.D. (chemistry); resigned, August 1968; Group 6, August 1967.

Lounge, John M. (June 28, 1946, Denver, Colo., but considers Burlington, Colo., his hometown)—Civilian; B.S. (physics and mathematics), M.S. (astrogeophysics); flight, mission specialist; Group 9, August 1981.

Lousma, Jack R. (Feb. 29, 1936, Grand Rapids, Mich.)—Colonel, USMC; B.S. (aeronautical engineering); flight, pilot; Group 5, April 1966; Skylab 3, commander STS-3, March 22–30, 1982. Resigned from NASA, 1984.

Lovell, James A., Jr. (March 25, 1928, Cleveland, Ohio)—Captain, USN (Ret.); B.S. from U.S. Naval Academy; Group 2, September 1962; Gemini 7, Gemini 12, Apollo 8, Apollo 13; served as Deputy Director of Science and Applications, Johnson Space Center, May 1971–March 1973; retired from Navy and resigned from NASA, March 1973. Group Vice President, Centel Corp., Chicago.

Lucid, Shannon W. (Jan. 14, 1943, Shanghai, China, but considers Bethany, Okla., her hometown)—Civilian; B.S. (chemistry), M.S. and Ph.D. (biochemistry); flight, mission specialist; Group 8, August 1979.

Mattingly, Thomas K., II (March 17, 1936, Chicago, Ill.)—Captain, USN; B.S. (aeronautical engineering); flight, pilot; Group 5, April 1966; Apollo 16; commander, STS-4, June 27–July 4, 1982.

McBride, Jon A. (Aug. 14, 1943, Charleston, W.Va., but considers Beckley, W.Va., his hometown)—Commander, USN; B.S. (aeronautical engineering); flight, pilot; Group 8, August 1979. Pilot, 41-G, Oct. 5–13, 1984.

McCandless, Bruce, II (June 8, 1937, Boston, Mass.)—Captain, USN; B.S. (naval sciences), M.S. (electrical engineering); flight, pilot; Group 5, April 1966; mission specialist, 41-B, Feb. 3–10, 1984.

McDivitt, James A. (June 10, 1929, Chicago, Ill.)—Brig. General, USAF (Ret.); B.S. (aeronautical engineering); Group 2, September 1962; Gemini 4, Apollo 9; was Manager, Apollo Spacecraft Program, Johnson Space Center, September 1969–1972; retired from Air Force and resigned from NASA, June 1972. Senior Vice President, Strategic Management, Rockwell International Corp., Pittsburgh, Pa.

McNair, Ronald E. (Oct. 21, 1950, Lake City, S.C.)—Civilian; B.S. and Ph.D. (physics); flight, mission specialist; Group 8, August 1979; mission specialist, 41-B, Feb. 3–10, 1984.

Michel, F. Curtis (June 5, 1934, LaCrosse, Wis.)—Civilian; B.S. and Ph.D. (physics); Group 4, June 1965; resigned from NASA, August 1969. Department of Space Physics and Astronomy, Rice University, Houston.

Mitchell, Edgar D. (Sept. 17, 1930, Hereford, Texas, but considers Artesia, N.M., his hometown)—Captain, USN (Ret.); B.S. (industrial management and aeronautical engineering), D.Sc. (aeronautics/astronautics); Group 5, April 1966; Apollo 14; retired from Navy and resigned from NASA, October 1972. Chairman of Board, Forecast Systems, Inc., West Palm Beach, Fla.

Mullane, Richard M. (Sept. 10, 1945, Wichita Falls, Texas, but considers Albuquerque, N.M., his hometown)—Lt. Colonel, USAF; B.S. (military engineering), M.S. (aeronautical engineering); flight, mission specialist; Group 8, August 1979; mission specialist, 41-D, Aug. 30–Sept. 5, 1984.

Musgrave, F. Story (Aug. 19, 1935, Boston, Mass., but considers Lexington, Ky., his hometown)—Civilian; B.S. (mathematics and statistics), B.A. (chemistry), M.B.A. (operations analysis and computer programming), M.D., M.S. (physiology and biophysics); flight, mission specialist; Group 6, August 1967; mission specialist on STS-6, April 4–9, 1983.

Nagel, Steven R. (Oct. 27, 1946, Canton, Ill.)—Major, USAF; B.S. (aeronautical and astronautical engineering), M.S. (mechanical engineering); flight, pilot; Group 8, August 1979.

Nelson, George. D. (July 13, 1950, Charles City, Iowa, but considers Willmar, Minn., his hometown)—B.S. (physics), M.S. and Ph.D. (astronomy); flight, mission specialist; Group 8, August 1979; mission specialist, 41-C, April 6–14, 1984.

O'Connor, Bryan D. (Sept. 6, 1946, Orange, Calif., but considers Twentynine Palms, Calif., his hometown)—Major, USMC; B.S. (engineering), M.S. (aeronautical systems); flight, pilot; Group 9, August 1981.

O'Leary, Brian T. (Jan. 27, 1949, Boston, Mass.)—Civilian; Ph.D. (astronomy); resigned, April 1968; Group 6, August 1967.

Onizuka, Ellison S. (June 24, 1946, Kealakekua, Kona, Hawaii)—Major, USAF; B.S. and M.S. (aerospace engineering); flight, mission specialist; Group 8, August 1979.

Overmyer, Robert F. (July 14, 1936, Lorain, Ohio, but considers Westlake, Ohio, his hometown)—Colonel, USMC; B.S. (physics), M.S. (aeronautics); flight, pilot; Group 7, August 1969; pilot for STS-5, Nov. 11–16, 1982; designated spacecraft commander for STS-18 (Spacelab 3).

Parker, Robert A. (Dec. 14, 1936, New York City, but considers Shrewsbury, Mass., his hometown)—Civilian; B.A. (astronomy and physics), Ph.D. (astronomy); flight, mission specialist; Group 6, August 1967; mission specialist for STS-9 (Spacelab 1). Nov. 28–Dec. 8, 1983.

Peterson, Donald H. (Oct. 22, 1933, Winona, Miss.)—Colonel, USAF (Ret.); B.S. from U.S. Military Academy, M.S. (nuclear engineering); flight, pilot; Group 7, August 1969; retired from Air Force, January 1980; mission specialist on STS-6, April 4–9, 1983.

Pogue, William R. (Jan. 23, 1930, Okemah, Okla.)—Colonel, USAF (Ret.); B.S. (education), M.S. (mathematics); Group 5, April 1966; Skylab 4; resigned from NASA, September 1975, and retired from Air Force, September 1975. Self-employed as a consultant to aerospace and energy firms.

Resnik, Judith A. (April 5, 1949, Akron, Ohio)—Civilian; B.S. and Ph.D. (electrical engineering); flight, mission specialist; Group 8, August 1979; mission specialist, 41-D, Aug. 30–Sept. 5, 1984.

Richards, Richard N. (Aug. 24, 1946, Key West, Fla., but considers St. Louis, Mo., his hometown)—Lt. Commander, USN; B.S. (chemical engineering), M.S. (aeonautical systems); flight, pilot; Group 9, August 1981.

Ride, Sally K. (May 26, 1951, Los Angeles, Calif.)—Civilian; B.A. (English), B.S., M.S. and Ph.D. (physics); flight, mission specialist; Group 8, August 1979; mission specialist on STS-7, June 18–24, 1983; mission specialist, 41-G, Oct. 5–13, 1984.

Roosa, Stuart A. (Aug. 15, 1933, Durango, Colo.)—Colonel, USAF (Ret.); B.S. (aeronautical engineering); Group 5, April 1966; Apollo 14; resigned from NASA and retired from Air Force, February 1976. President and owner, Gulf Coast Coors, Inc., Gulfport, Miss.

Ross, Jerry L. (Jan. 20, 1948, Crown Point, Ind.)—Captain, USAF; B.S. and M.S. (mechanical engineering); flight, mission specialist; Group 9, August 1981.

Schirra, Walter M., Jr. (March 12, 1923, Hackensack, N.J.)—Captain, USN (Ret.); B.S. from U.S. Naval Academy; Group 1, April 1959; Mercury 8, Gemini 6, Apollo 7; resigned from NASA and retired from Navy, July 1969. President, Schirra Enterprises.

Schmitt, Harrison H. (July 3, 1935, Santa Rita, N.M.)—Civilian; B.S. (science), Ph.D. (geology); Group 4, June 1965; Apollo 17; Special Assistant to NASA Administrator for Energy Research and Development, February 1974; appointed NASA Assistant Administrator for Energy Programs, May 1974; resigned from NASA, August 1975. Elected U.S. Senator from New Mexico in November 1976; defeated for reelection in November 1982.

Schweickart, Russell L. (Oct. 25, 1935, Neptune, N.J.)—Civilian; B.S. (aeronautical engineering), M.S. (aeronautics and astronautics); Group 3, October 1963; Apollo 9; transferred to NASA Headquarters, Washington, D.C., May 1, 1974; detailed to California Governor in 1977 under Intergovernmental Personnel Act; resigned from NASA, August 1979. Chairman, California Energy Commission, Sacramento, Calif.

Scobee, Francis R. (May 19, 1939, Cle Elum, Wash.)—Major, USAF, (Ret.); B.S. (aerospace engineering); flight, pilot; Group 8, August 1979; retired from Air Force, January 1980; pilot, 41-C, April 6–14, 1984.

Scott, David R. (June 6, 1932, San Antonio, Texas)—Colonel, USAF (Ret.); B.S. from U.S. Military Academy, M.S. (aeronautics and astronautics); Group 3, October 1963; Gemini 8, Apollo 9, Apollo 15; Special Assistant for Mission Operations, Apollo Spacecraft Program Office, Johnson Space Center, July 1972–August 1973; Deputy Director, Dryden Flight Research Center, Edwards, Calif., August 1973–April 1975; appointed Center Director 1975; resigned from NASA, October 1977. President, Scott Science & Technology, Inc., Lancaster, Calif.

Seddon, Margaret R. (Nov. 8, 1947, Murfreesboro, Tenn.)—Civilian; B.A. (physiology), M.D.; flight, mission specialist; Group 8, August 1979.

See, Elliott M., Jr. (July 23, 1927, Dallas, Texas)—Civilian; B.S. from U.S. Merchant Marine Academy, M.S. (engineering); deceased; Group 2, September 1962; died in T-38 crash with Charles Bassett, Feb. 28, 1966, Lambert Municipal Airport, St. Louis.

Shaw, Brewster H., Jr. (May 16, 1945, Cass City, Mich.)—Major, USAF; B.S. and M.S. (engineering mechanics); flight, pilot; Group 8, August 1979; pilot, STS-9 (Spacelab 1), Nov. 28–Dec. 8, 1983.

Shepard, Alan B., Jr. (Nov. 18, 1923, East Derry, N.H.)—Rear Admiral, USN (Ret.); B.S. from U.S. Naval Academy; Group 1, April 1959; Mercury 3, Apollo 14; resigned from NASA and retired from Navy, August 1974. President, Windward Co., Deer Park, Texas.

Shriver, Loren J. (Sept. 23, 1944, Jefferson, Iowa, but considers Paton, Iowa, his hometown)—Major, USAF; B.S. (aeronautical engineering), M.S. (astronautical engineering); flight, pilot; Group 8, August 1979.

Slayton, Donald K. (March 1, 1924, Sparta, Wis.)—Major, USAF (Reserve); B.S. (aeronautical engineering); Group 1, April 1959; Apollo-Soyuz Test Project; was Manager for Orbital Flight Tests, Space Shuttle Program Office, Johnson Space Center; retired from NASA, February 1982. President, Space Services, Inc., and a consultant to aerospace corporations.

Smith, Michael J. (April 30, 1945, Morehead City, N.C.)—Commander, USN; B.S. (naval science), M.S. (aeronautical engineering); flight, pilot; Group 9, August 1981.

Spring, Sherwood C. (Sept. 3, 1944, Hartford, Conn., but considers Harmony, R.I., his hometown)—Major, USA; B.S. (general engineering), M.S. (aerospace engineering); flight, mission specialist; Group 9, August 1981.

Springer, Robert C. (May 21, 1942, St. Louis, Mo., but considers Ashland, Ohio, his hometown)—Lt. Colonel, USMC; B.S. (naval science), M.S. (operations research and systems analysis); flight, mission specialist; Group 9, August 1981.

Stafford, Thomas P. (Sept. 17, 1930, Weatherford, Okla.)—Lt. General, USAF (Ret.); B.S. from U.S. Naval Academy; Group 2, September 1962; Gemini 6, Gemini 9, Apollo 10, Apollo-Soyuz Test Project; resigned from NASA, November 1975, and retired from Air Force, November 1979. Vice Chairman, Gibraltar Exploration, Ltd., Oklahoma City.

Stewart, Robert L. (Aug. 13, 1942, Washington, D.C., but considers Arlington, Texas, his hometown)—Lt. Colonel, USAF: B.S. (mathematics), M.S. (aerospace engineering); flight, mission specialist; Group 8, August 1979; mission specialist, 41-B, Feb. 3–10, 1984.

Sullivan, Kathryn D. (Oct. 3, 1951, Paterson, N.J., but considers Woodland Hills, Calif., her hometown)—Civilian, B.S. (earth sciences), Ph.D. (geology); flight, mission specialist; Group 8, August 1979. Mission specialist, 41-G, Oct. 5–13, 1984.

Swigert, John L., Jr. (Aug. 30, 1931, Denver, Colo.)—Civilian; B.S. (mechanical engineering), M.S. (aerospace science), M.B.A.; Group 5, April 1966; Apollo 13; resigned from NASA, July 1978. Staff Director, Committee on Science and Astronautics, House of Representatives, April 1973–September 1977. In November 1982, won the new seat for Colorado's Sixth Congressional District; died of complications from cancer in Washington, D.C., Dec. 27, 1982, a week before he would have taken the congressional seat.

Thagard, Norman E. (July 3, 1943, Marianna, Fla., but considers Jacksonville, Fla., his hometown)—Civilian; B.S. and M.S. (engineering science), M.D.; flight, mission specialist; Group 8, August 1979; mission specialist on STS-7, June 18–24, 1983; designated a mission specialist for STS-18 (Spacelab 3).

Thornton, William E. (April 14, 1929, Faison, N.C.)—Civilian; B.S. (physics), M.D.; flight, mission specialist; Group 6, August 1967; mission specialist STS-8, Sept. 30–Oct. 5, 1983.

Truly, Richard H. (Nov. 12, 1937, Fayette, Miss.)—Captain, USN; B.S. (aeronautical engineering); flight, pilot; Group 7, August 1969; piloted *Enterprise* space shuttle free-flight approach and landing tests 2 and 4 on Sept. 13 and Oct. 12, 1977; pilot on STS-2, Nov. 12–14, 1981; commander STS-8, Sept. 30–Oct. 5, 1983. Resigned, 1984.

van Hoften, James D. (June 11, 1944, Fresno, Calif., but considers Burlingame, Calif., his hometown)—Civilian; B.S. (civil engineering), M.S. (hydraulic engineering), Ph.D. (fluid mechanics); flight, mission specialist; Group 8, August 1979; mission specialist, 41-C, April 6–14, 1984.

Walker, David M. (May 20, 1944, Columbus, Ga., but considers Eustis, Fla., his hometown)—Commander, USN; B.S. from U.S. Naval Academy; flight, pilot; Group 8, August 1979. Pilot, 51-A, Nov. 8–16, 1984.

Weitz, Paul J. (July 25, 1932, Erie, Pa.)—Captain, USN (Ret.); B.S. and M.S. (aeronautical engineering); flight, pilot; Group 5, April 1966; Skylab 2; retired from Navy, June 1976; commander of STS-6, April 4–9, 1983.

White, Edward H., II (Nov. 14, 1930, San Antonio, Texas)—Lt. Colonel, USAF; B.S. from U.S. Military Academy, M.S. (aeronautical engineering); deceased; Group 2, September 1962; Gemini 4; died in Apollo spacecraft fire at Kennedy Space Center, Jan. 27, 1967.

Williams, Cifton C., Jr. (Sept. 26, 1932, Mobile, Ala.)—Major, USMC; B.S. (mechanical engineering); deceased; Group 3, October 1963; died in T-38 crash near Tallahassee, Fla., Oct. 5, 1967.

Williams, Donald E. (Feb. 13, 1942, Lafayette, Ind.)—Commander, USN; B.S. (mechanical engineering); flight, pilot; Group 8, August 1979.

Worden, Alfred M. (Feb. 7, 1932, Jackson, Mich.)—Colonel, USAF (Ret.); B.S. (military science) from U.S. Military Academy, M.S. (astronautical and aeronautical engineering and instrumentation engineering); Group 5, April 1966; Apollo 15; 1972–1973, Senior Aerospace Scientist, Ames Research Center, Mountain View, Calif.; 1973–1975, Chief, Systems Studies Division at Ames; resigned from Air Force and NASA, September 1975. President, Alfred M. Worden, Inc., Palm Beach Gardens, Fla.

Young, John W. (Sept. 24, 1930, San Francisco, Calif.)—Captain, USN (Ret.); B.S. (aeronautical engineering); flight, pilot; Group 2, September 1962; retired from Navy, September 1976; Gemini 3, Gemini 10, Apollo 10, Apollo 16, STS-1; Chief, Astronaut Office, Johnson Space Center, Houston; commander on STS-1, April 12–14, 1981; commander of STS-9 (Spacelab 1).

Q *Who was the first space traveler to receive a promotion for a spaceflight?*

A Lt. Yuri Gagarin, who became Maj. Gagarin in orbit. The promotion, coupled with Soviet secrecy, caused some confusion back home. Even after Gagarin had landed, his father refused to believe that his son had been the space traveler, because "Our Yuri is only a lieutenant" and Gagarin is a common name.

ASTRONAUT GROUP SELECTIONS

Since April 1959, NASA has selected 127 pilots and scientists in nine different groups. Through December 1984, 86 of them have flown on 93 missions. Here is the breakdown of each group and the date it was selected. Flight: 73; retired: 1; resigned: 44; deceased: 9.

Group 1: April 1959

NAME	STATUS
Carpenter, M. Scott	Resigned
Cooper, L. Gordon	Resigned
Glenn, John H., Jr.	Resigned
Grissom, Virgil I.	Deceased
Schirra, Walter M., Jr.	Resigned
Shepard, Alan B., Jr.	Resigned
Slayton, Donald K.	Retired

Group 2: September 1962

NAME	STATUS
Armstrong, Neil A.	Resigned
Borman, Frank	Resigned
Conrad, Charles, Jr.	Resigned
Lovell, James A., Jr.	Resigned
McDivitt, James A.	Resigned
See, Elliott M., Jr.	Deceased
Stafford, Thomas P.	Resigned
White, Edward H., II	Deceased
Young, John W.	Flight

Group 3: October 1963

NAME	STATUS
Aldrin, Edwin E., Jr.	Resigned
Anders, William A.	Resigned
Bassett, Charles A.	Deceased
Bean, Alan L.	Resigned
Cernan, Eugene A.	Resigned
Chaffee, Roger B.	Deceased
Collins, Michael	Resigned
Cunningham, Walter	Resigned
Eisele, Donn F.	Resigned
Freeman, Theodore C.	Deceased
Gordon, Richard F., Jr.	Resigned
Schweickart, Russell L.	Resigned
Scott, David R.	Resigned
Williams, Clifton C., Jr.	Deceased

Group 4: June 1965

NAME	STATUS
Garriott, Owen K.	Flight
Gibson, Edward G.	Resigned
Graveline, Duane E.	Resigned
Kerwin, Joseph P.	Flight
Michel, F. Curtis	Resigned
Schmitt, Harrison H.	Resigned

Group 5: April 1966

NAME	STATUS
Brand, Vance D.	Flight
Bull, John S.	Resigned
Carr, Gerald P.	Resigned
Duke, Charles M., Jr.	Resigned
Engle, Joe H.	Flight
Evans, Ronald E.	Resigned
Givens, Edward G.	Deceased
Haise, Fred W., Jr.	Resigned
Irwin, James B.	Resigned
Lind, Don L.	Flight
Lousma, Jack R.	Resigned
Mattingly, Thomas K., II	Flight
McCandless, Bruce, II	Flight
Mitchell, Edgar D.	Resigned
Pogue, William R.	Resigned
Roosa, Stuart A.	Resigned
Swigert, John L., Jr.	Deceased
Weitz, Paul J.	Flight
Worden, Alfred M.	Resigned

Group 6: August 1967

NAME	STATUS
Allen, Joseph P.	Flight
Chapman, Philip K.	Resigned
England, Anthony W.	Flight
Henize, Karl G.	Flight
Holmquest, Donald L.	Resigned
Lenoir, William B.	Resigned
Llewellyn, John A.	Resigned*
Musgrave, F. Story	Flight
O'Leary, Brian T.	Resigned*
Parker, Robert A.	Flight
Thornton, William E.	Flight

Group 7: August 1969

NAME	STATUS
Bobko, Karol J.	Flight
Crippen, Robert L.	Flight
Fullerton, C. Gordon	Flight
Hartsfield, Henry W., Jr.	Flight
Overmyer, Robert F.	Resigned
Peterson, Donald H.	Flight
Truly, Richard H.	Resigned

* Resigned before achieving flight status.

Group 8: January 1978

NAME	STATUS	NAME	STATUS
Bluford, Guion S., Jr.	Flight	Stewart, Robert L.	Flight
Brandenstein, Daniel C.	Flight	Sullivan, Kathryn D.	Flight
Buchli, James F.	Flight	Thagard, Norman E.	Flight
Coats, Michael L.	Flight	van Hoften, James D.	Flight
Covey, Richard O.	Flight	Walker, David M.	Flight
Creighton, John O.	Flight	Williams, Donald E.	Flight
Fabian, John M.	Flight		
Fisher, Anna L.	Flight		
Gardner, Dale A.	Flight		
Gibson, Robert L.	Flight	**Group 9: July 1980**	
Gregory, Frederick D.	Flight	Bagian, James P.	Flight
Griggs, S. David	Flight	Blaha, John E.	Flight
Hart, Terry J.	Resigned	Bolden, Charles F., Jr.	Flight
Hauck, Frederick H.	Flight	Bridges, Roy D., Jr.	Flight
Hawley, Steven A.	Flight	Chang, Franklin R.	Flight
Hoffman, Jeffrey A.	Flight	Cleave, Mary L.	Flight
Lucid, Shannon W.	Flight	Dunbar, Bonnie J.	Flight
McBride, Jon A.	Flight	Fisher, William F.	Flight
McNair, Ronald E.	Flight	Gardner, Guy S.	Flight
Mullane, Richard M.	Flight	Grabe, Ronald J.	Flight
Nagel, Steven R.	Flight	Hilmers, David C.	Flight
Nelson, George D.	Flight	Leestma, David C.	Flight
Onizuka, Ellison S.	Flight	Lounge, John M.	Flight
Resnik, Judith A.	Flight	O'Connor, Bryan D.	Flight
Ride, Sally K.	Flight	Richards, Richard N.	Flight
Scobee, Francis R.	Flight	Ross, Jerry L.	Flight
Seddon, Margaret R.	Flight	Smith, Michael J.	Flight
Shaw, Brewster H., Jr.	Flight	Spring, Sherwood C.	Flight
Shriver, Loren J.	Flight	Springer, Robert C.	Flight

Q *When was NASA established?*

A The National Aeronautics and Space Administration replaced the forty-three-year-old National Advisory Committee for Aeronautics in August 1958.

Q *Which was the only U.S. manned space program that did not have to get formal approval from Congress prior to becoming official, and who named it?*

A Gemini, because it had originally been included in appropriations for Project Mercury and named Mercury Mark II. NASA's Alex Nagy named it, continuing the custom of using terms from classical mythology.

Q *Who selects the crews for shuttle missions?*

A The director of Flight Operations, currently George Abbey, and the chief of the Astronaut Office, currently John Young. The selection process itself is one of NASA's best-kept secrets.

Project Mercury MANNED MISSIONS. Of the Original 7 Astronauts, six would fly during Project Mercury; one would later die in a launch test fire, another would participate in both the Gemini and Apollo programs, and still another would eventually walk on the Moon. The only one of the Original 7 not to fly in Project Mercury, Deke Slayton, would later spend more time in space during the Apollo-Soyuz mission than the combined time of all the Project Mercury flights. What follows is mission data for the six flights in the program. MR designations are for Mercury-Redstone; MA represents Mercury-Atlas.

MISSION	ASTRONAUT	AGE	SPACE-CRAFT	BACKUP	DATE	LAUNCH TIME	DURATION*	ORBITS	RECOVERY	RECOVERY SHIP
MR-3 *Freedom 7*	Shepard	37	7	Glenn	5/5/61	09:34	0:15:28	—	Atlantic	USS *Lake Champlain*
MR-4 *Liberty Bell 7*	Grissom	35	11	Glenn	7/21/61	07:20	0:15:37	—	Atlantic	USS *Randolph*
MA-6 *Friendship 7*	Glenn	30	13	Carpenter	2/20/62	09:47:39	4:55:23	3	Atlantic	USS *Noa*
MA-7 *Aurora 7*	Carpenter	37	18	Schirra	5/24/62	07:45:16	4:56:05	3	Atlantic	USS *Intrepid*
MA-8 *Sigma 7*	Schirra	39	16	Cooper	10/3/62	07:15:11	9:13:11	6	Pacific	USS *Kearsarge*
MA-9 *Faith 7*	Cooper	36	20	Shepard	5/15/63	08:04:13	34:19:49	22.5	Pacific	USS *Kearsarge*

* Indicates hours:minutes:seconds.

Project Mercury UNMANNED TESTS. Before Alan Shepard climbed into a spacecraft on May 5, 1961, and became the first American in space, NASA conducted no fewer than seventeen unmanned tests of the spacecraft systems and launch vehicles. Listed below are all the various tests—which took place between August 21, 1959, and November 29, 1961—and their data, objectives, and results.

MISSION	DATE	SPACE-CRAFT	DURATION	OCCUPANT	OBJECTIVE	RESULT
Little Joe 1	8/21/59	BP	0:00:20	—	Max abort to test launch escape system and recovery	F
Big Joe	9/9/59	BP	0:13:00	—	Structures, heat protection, and recovery systems	S/F
Little Joe 6	10/4/59	BP	0:05:10	—	Launch vehicle qualification and capsule evaluation	P
Little Joe 1A	11/4/59	BP	0:08:11	—	Repeat of Little Joe 1	P
Little Joe 2	12/4/59	BP	0:11:06	RM (Sam)	Abort with rhesus monkey	S
Little Joe 1B	1/21/60	BP	0:08:35	RM (Miss Sam)	Repeat of Little Joe 1A	S
Beach Abort	5/9/60	1	0:01:16	—	Pad abort to qualify spacecraft	S
Mercury-Atlas 1	7/29/60	4	0:03:18	—	Spacecraft launch vehicle compatibility in abort	F
Little Joe 5	11/8/60	3	0:02:22	—	Max abort to qualify launch escape system	F
Mercury-Redstone 1	11/21/60	2	0:00:02	—	Qualification of spacecraft and launch vehicle	F
Mercury-Redstone 1A	12/19/60	2A	0:15:45	—	Repeat of Mercury-Redstone 1	S
Mercury-Redstone 2	1/31/61	5	0:16:39	C (Ham)	Suborbital abort with chimpanzee	S/P
Mercury-Atlas 2	2/21/61	6	0:17:56	—	Repeat of Mercury-Atlas 1	S
Little Joe 5A	3/18/61	14	0:23:48	—	Repeat of Little Joe 5	P
Mercury-Redstone Booster Development	3/24/61	BP	0:08:23	—	Suborbital test of Redstone	S
Mercury-Atlas 3	4/25/61	8	0:07:19	Simulated man	Launch vehicle, spacecraft, and tracking	F
Little Joe 5B	4/28/61	14A	0:05:25	—	Repeat of Little Joe 5A	S/P
Mercury-Atlas 4	9/13/61	8A	1:49:20	Simulated man	Repeat of Mercury-Atlas 3 and environmental system check	P
Mercury-Scout	11/1/61	—	0:00:43	—	Tracking checks	F
Mercury-Atlas 5	11/29/61	9	3:20:59	C (Enos)	Life-support functions with chimp	P/S

PROJECT MERCURY

Characteristics of the Spacecraft

What exactly was a Mercury spacecraft? How long was it? How heavy? What kind of an environment did it provide for the astronauts? The information below pertains to the spacecrafts flown by John Glenn and Scott Carpenter during their 3-orbit missions. There were some modifications for the last two flights with Wally Schirra and Gordon Cooper, and that information appears in parentheses where applicable.

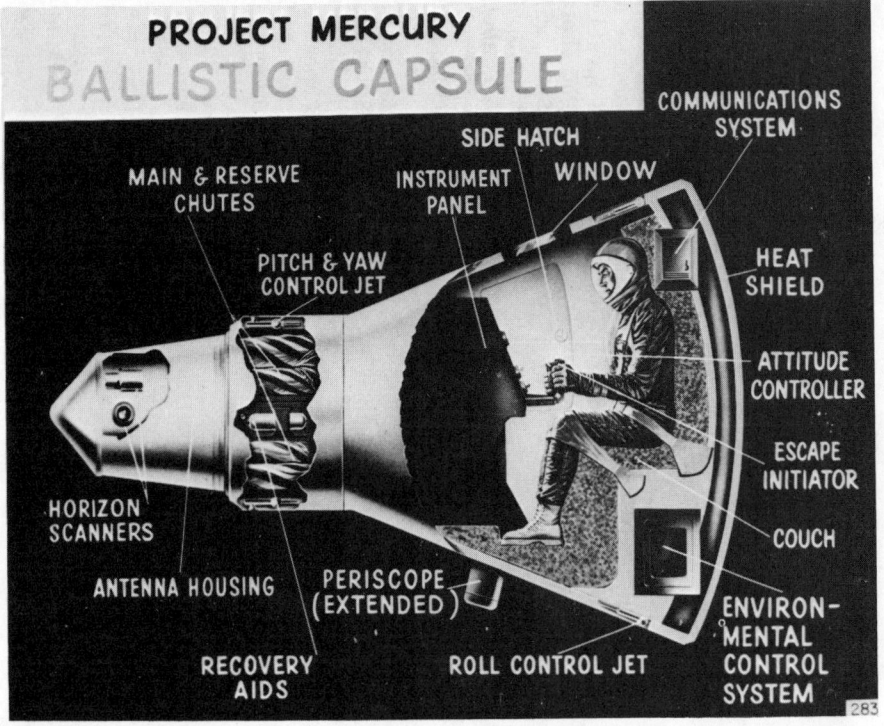

PHYSICAL DATA

 Length: 7.91 m on launchpad
     3.34 m without escape rocket
     2.92 m without retro-package

 Diameter: 1.89 m bottom
      81.3 cm top

 Weight: 1,935 kg launch
     1,355 kg orbit
     1,131 kg splashdown
     1,099 kg recovery

ENVIRONMENTAL CONTROL

Atmosphere: 100% oxygen at nominal 258 mmHg

Storage: 2 bottles totaling 3.6 kg (MA-9: 3 bottles totaling 5.4 kg)

Cooling fluid: 17.7 kg water (MA-9: 21.8 kg)

Drinking water: 2.5 kg (MA-9: 4.5 kg)

Atmosphere cleanser: 2 kg lithium hydroxide (MA-8, MA-9: 2.4 kg)

Odor absorber: 0.45 kg activated charcoal (MA-9: 0.091 kg)

ELECTRICAL SYSTEMS

Batteries: 3 × 3.000 watt-hr silver zinc on main 24 volt DC buses
2 × 1.500 watt-hr silver zinc on 24, 18, 12.8, and 6 volt
DC standby buses (MA-9: 2 × 3.000 watt-hr batteries)

ATTITUDE CONTROL THRUSTERS

	SYSTEM A (KG THRUST)	SYSTEM B (KG THRUST)	NO. OF THRUSTERS
Pitch	10.9	1.8–10.9	6
Roll	2.7	0.4–2.7	6
Yaw	10.9	1.8–10.9	6

PROPELLANT

Automotive (ASCS/FBW)	15.8 kg
Manual (MP/RSCS)	11.3 kg
Total	27.1 kg (MA-9: 33.9 kg) hydrogen peroxide

ATTITUDE CONTROL MODES*

	ATTITUDE THRUSTER SYSTEM	ELECTRICAL POWER
ASCS	A	DC and AC
FBW	A	DC
MP	B	none
RSCS	B	DC and AC

SOLID ROCKET MOTORS

	NUMBER OF MOTORS	NOMINAL THRUST (KG)	BURN TIME (SEC)
Escape	1	23.587	1
Tower jettison	1	363	1.5
Posigrade	3	181	1
Retrograde	3	454	10

PARACHUTES

Drogue: 1 × 1.8 m conical ribbon

Main: 1 × 19.2 m ringsail plus identical reserve

* ASCS: Attitude Stabilization & Control System
FBW: Fly-By-Wire
MP: Manual Proportional
RSCS: Rate Stabilization & Control System (not installed before MR-4 or
for MA-9)

206

Project Gemini MANNED MISSIONS. There were ten manned missions in Project Gemini. Of the twenty astronauts who flew, six would eventually walk on the Moon in Project Apollo. Only one spacecraft had a name, Gus Grissom and John Young's Gemini 3, *Molly Brown*.

MISSION	ASTRONAUT	AGE	SPACE-CRAFT	BACKUP	DATE	LAUNCH TIME	DURATION	RECOVERY	RECOVERY SHIP
3	Grissom, Young	38, 34	3	Schirra, Stafford	3/23/65	09:24:00.064	4:52:31	Atlantic	USS *Intrepid*
4	McDivitt, White	35, 34	4	Borman, Lovell	6/3/65	10:15:59.562	97:56:12	Atlantic	USS *Wasp*
5	Cooper, Conrad	38, 35	5	Armstrong, See	8/21/65	08:59:59.518	190:55:14	Atlantic	USS *Lake Champlain*
6-A	Schirra, Stafford	42, 35	6	Grissom, Young	12/15/65	08:37:26.471	25:51:54	Atlantic	USS *Wasp*
7	Borman, Lovell	37, 37	7	White, Collins	12/4/65	14:30:03.702	330:35:01	Atlantic	USS *Wasp*
8	Armstrong, Scott	35, 33	8	Conrad, Gordon	3/16/66	11:41:02.389	10:41:26	Pacific	USS *Mason*
9-A	Stafford, Cernan	35, 32	9	Lovell, Aldrin	6/3/66	08:39:33.335	72:20:50	Atlantic	USS *Wasp*
10	Young, Collins	35, 35	10	Bean, Williams	7/18/66	17:20:26.648	70:46:39	Atlantic	USS *Guadalcanal*
11	Conrad, Gordon	36, 36	11	Armstrong, Anders	9/12/66	09:42:26.546	71:17:08	Atlantic	USS *Guam*
12	Lovell, Aldrin	38, 36	12	Cooper, Cernan	11/11/66	15:46:33.419	94:34:31	Atlantic	USS *Wasp*

APPENDIXES

Project Gemini UNMANNED TESTS. Unlike the seventeen unmanned tests conducted before a human was launched in the Mercury program, the Gemini unmanned tests were limited to two. This was partially due to an ability to accept information previously collected in the earlier program. By the time Gemini began to count down for a two-man launch, NASA was satisfied that the majority of systems would operate as expected and needed only to qualify the spacecraft and launcher and test the suborbital systems.

MISSION	DATE	SPACE-CRAFT	OCCUPANT	RECOVERY SHIP	OBJECTIVE	RESULT*
1	4/8/64	1	—	—	Spacecraft/launcher qualification	S
2	1/19/65	2	Simulated man	USS *Lake Champlain*	Suborbital systems test	S

* S: Success

PROJECT GEMINI

Characteristics of the Spacecraft

The two-man Gemini spacecraft were considerably more complex than their one-man Mercury predecessors. The information below describes a typical spacecraft.

SIZE	ADAPTER	RE-ENTRY MODULE	TOTAL
Length	2.286 m	3.657 m	5.736 m
Diameter, minimum	2.286 m	98.2 cm	
Diameter, maximum	3.048 m	2.286 m	

WEIGHT

3,400 kg orbit
2,200 kg splashdown

ENVIRONMENTAL CONTROL

Atmosphere: 100% oxygen at nominal 258 mmHg

Temperature: 24°C (49°C during re-entry)

Suit circuit: 100% oxygen at 195–217 mmHg

Storage: 1 Adapter sphere (47.17 kg) prime containing supercritical oxygen
2 Re-entry Module bottles secondary containing gaseous oxygen

Drinking water: 19.1 kg capacity tank in Adapter
nominal 6.6 kg capacity in Re-entry Module

Fuel cell product water: nominal 19.1 kg capacity tank

Atmosphere cleaner: lithium hydroxide

Odor absorber: charcoal

ELECTRICAL SYSTEMS
Main: 2 fuel cells or combinations of 400 ampere-hr silver zinc batteries
Secondary: 4 × 45 ampere-hr silver zinc batteries during re-entry
3 × 15 ampere-hr silver zinc squib batteries for pyrotechnics

Fuel cells (each):
Weight: 30.9 kg
Production: 1kw DC at 23.3–26.5 volts
Size: cylinder 61 cm by 30.5 cm diameter
No. of stacks: 3
No. of cells per stack: 32
Reactants: Hydrogen and oxygen stored as liquids in separate tanks

ATTITUDE CONTROL THRUSTERS*
OAMS: 8 × 11.3 kg thrust engines attached to Adapter
RCS: 16 × 11.3 kg thrust engines on Re-entry Module for re-entry only

TRANSLATION CONTROL THRUSTERS*
OAMS: 6 × 45.4 kg thrust engines on Adapter
2 × 38.6 kg thrust engines on Adapter

RETROROCKETS
4 × 32.3 cm diameter solid propellant rockets fired sequentially for 5.5 sec each

PARACHUTES
Drogue: 1 × 3.27 m conical ribbon
Pilot: 1 × 5.58 m ringsail
Main: 1 × 25.6 m ringsail

* OAMS: Orbit Attitude Maneuvering System
RCS: Re-entry Control System

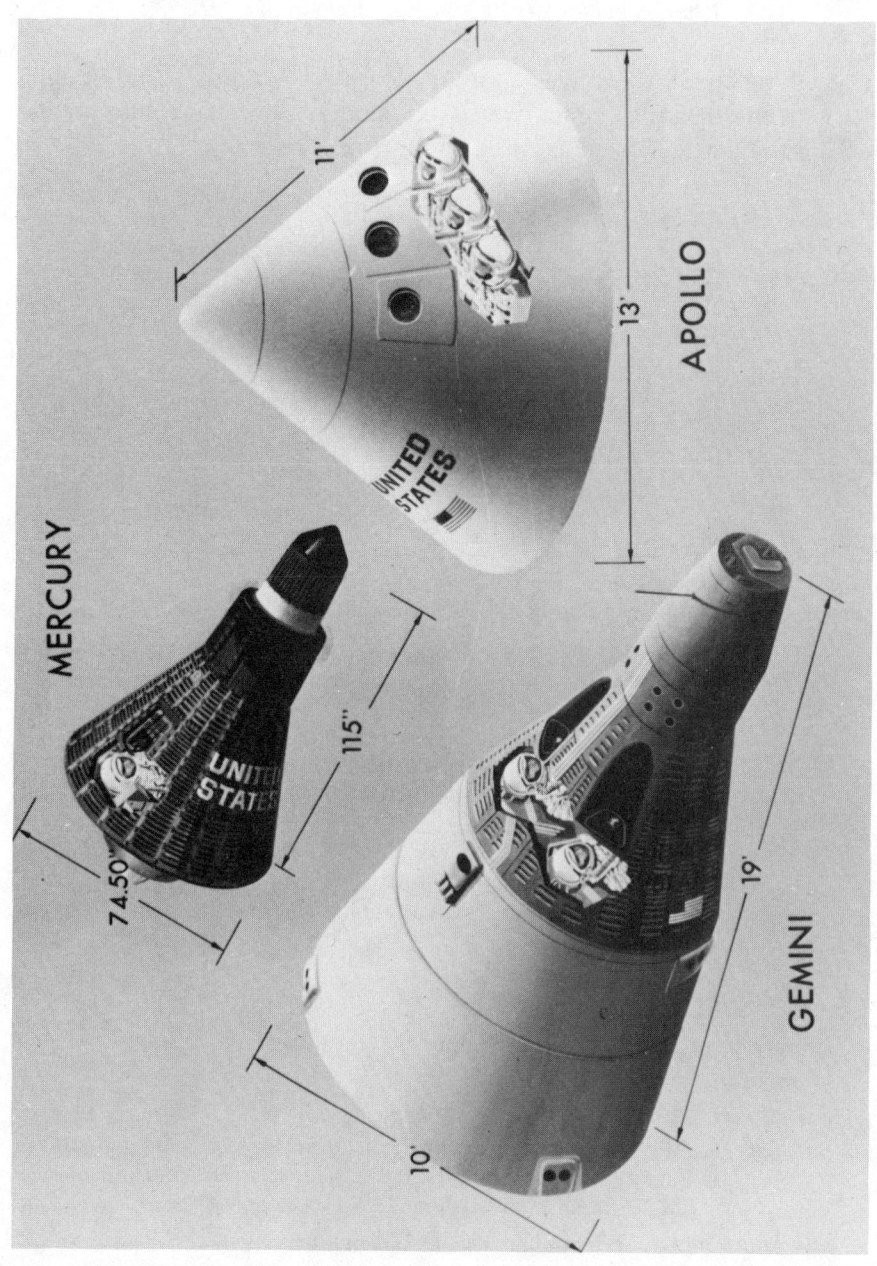

PROJECT APOLLO

Characteristics of the Spacecraft

Everything about Apollo was gigantic in relation to Mercury and Gemini. As with the data on Gemini, we have included all available information except the particulars concerning liquid propellants.

PHYSICAL DATA*
  Length: SM: 4.52 m
            SM: with SPS nozzle 7.49 m
            CM: 3.22 m
            Total in-flight: 10.1 m
            LES: 10.06 m
            SLA: 8.53 m
            Total SLA-CSM-LES: 25 m

  Diameter: SM: 3.91 m
              CM: 3.91 m
              SLA top-bottom: 3.91/6.6 m

  Weight (in metric tons)   SM loaded: 24
                            CM loaded: 5.6
                            Splashdown (CM): 5.3
                            Total loaded: 29.6
                            LES: 4
                            SLA: 1.8
                            Total atop launcher plus LM: 52
                            Total CSM-LM orbital (typical): 46.2

ENVIRONMENTAL CONTROL
  Atmosphere: 100% oxygen at nominal 258 mmHg
  Temperature: 21.1° C–23.8° C
  Suit circuit: 100% oxygen at 190 mmHg
  Storage: 2 SM tanks containing 290 kg of liquid oxygen, of which 100 kg
             is for environmental control, the balance to fuel cells
             1 CM surge tank containing 3.7 kg gaseous oxygen for re-entry
  Drinking water: 15 liters capacity in CM from fuel cells
  Waste water: 26.5 liters capacity in CM from suit heat exchanger
  Atmosphere cleanser: lithium hydroxide
  Odor absorber: activated charcoal

ELECTRICAL SYSTEMS
> *Main*:   3 fuel cells
>
> *Secondary*:   3 × 40 ampere-hr silver zinc batteries during descent
> 2 × 0.75 ampere-hr silver zinc batteries for pyrotechnics
>
> *Fuel cells* (each):
> Weight: 111 kg
> Production: 1.4 kw DC at 25–30 volts
> Size: cylinder 1.1 m by 55.9 cm diameter
> No. of stacks: 1
> No. of cells: 31
>
> *Reactants*:   2 × 12.7 kg capacity hydrogen tanks at 15.8 kg/cm^2 (minimum)
> 2 × 145 kg capacity oxygen tanks at 60.8 kg/cm^2 (minimum)

ATTITUDE CONTROL THRUSTERS*
> *SMRCS*:   16 × 45.5 kg thrust engines on 4 SM quads
>
> *CMRCS*:   12 × 42.2 kg thrust engines around CM

TRANSLATION CONTROL
> SMRCS fired in groups of 2 or 4

SERVICE PROPULSION SYSTEM
> 1 × 9.3 thrust liquid propellant motor

LES SOLID ROCKET MOTORS

	LENGTH	DIAMETER	THRUST	BURN TIME	WEIGHT
Escape	4.71 m	66 cm	66.7 MT	3 sec	2.132 kg
Pitch control	55.9 cm	22.3 cm	1.1 MT	0.5 sec	22.7 kg
Tower jettison	1.41 m	65.4 cm	14.3 MT	1 sec	—

PARACHUTES
> *Drogue*:   2 × 5 m conical ribbon
>
> *Pilot*:   3 × 2.2 m ringshot
>
> *Main*:   3 × 25.45 m ringsail

* SPS: Service Propulsion System
LES: Launch Escape System
SLA: Spacecraft-LM Adapter
LM: Lunar Module
CSM: Command and Service Modules (CM, SM)
RCS: Reaction Control System

PROJECT APOLLO

Lunar Module

Figures in parentheses refer to the modified LM for Apollos 15–17 (1-series mission).

PHYSICAL DATA

	DESCENT STAGE	ASCENT STAGE
Height	3.2 m	3.76 m
Diameter	4.2 m	4.2 m
Total diameter (diagonally across legs)	9.4 m	
Dry weight	1.860 kg (2.760)	2.040 kg (2.130)
Loaded weight	10.000 kg (11.610)	4.670 kg (4.760)
Launch weight	14.670 kg (16.370)	

ENVIRONMENTAL CONTROL

Atmosphere:　100% oxygen at nominal 250 mmHg

Temperature:　23.9° C

Suit circuit:　100% oxygen at 190 mmHg

Storage:　1 (2) descent stage tank(s) containing 21.7 kg (43.5 kg) gaseous oxygen fitted to Quad 3 (3 and 4) at 192 kg/cm^2 (189 kg/cm^2) 2 ascent stage tanks, each containing 1.1 kg gaseous oxygen at 60 kg/cm^2

No. of cabin repressurization:　4

Water:　1 (2) × 151 kg capacity descent stage tank(s) fitted to Quad 2 (2 and 4) filled to 75% nitrogen pressurized 2 × 19.27 kg capacity ascent stage tanks

Atmosphere cleanser:　lithium hydroxide

Odor absorber:　activated charcoal

Coolant:　11.3 kg of 35% ethylene glycol, 65% water solution

ELECTRICAL SYSTEMS

Main: 4(5) × 400 ampere-hr (415 ampere-hr) silver zinc batteries in forward descent stage box structure (rear descent stage box structure)

Secondary: 2 × 296 ampere-hr silver zinc batteries in ascent stage

Power: 26–32 volt DC buses

ATTITUDE CONTROL THRUSTERS*

Ascent stage RCS: 16 × 45.4 kg thrust engines on 4 quads

TRANSLATION CONTROL

RCS: fired in groups of 2 or 4

MAIN PROPULSION*

1 × throttleable 476 kg to 4.477 (580 kg to 4.490 kg) thrust liquid propellant motor: DPS

1 × fixed 1.588 kg thrust liquid propellant motor: APS

LIQUID PROPELLANT

RCS quantity: Nominal 287 kg (286 kg) N_2O_4/Aerozene 50 in 2 oxidizer/2 fuel tanks

Pressurant: 0.93 kg helium stored at 214 kg/cm^2 in 2 tanks

DPS quantity: Nominal 8.187 kg (8.838 kg) N_2O_4/Aerozene 50 in 2 oxidizer/ 2 fuel tanks

Pressurant: 1 × supercritical helium tank containing 22 kg (23.2 kg) at 109.3 kg/cm^2 and +240° C

APS quantity: Nominal 2.353 kg (2.371 kg) N_2O_4/Aerozene 50 in 1 oxidizer/ 1 fuel tank

Pressurant: 2 × 2.9 kg helium tanks pressurized to 214 kg/cm^2

* RCS: Reaction Control System
 DPS: Descent Propulsion System
 APS: Ascent Propulsion System

Project Apollo UNMANNED TESTS. While the step from Mercury to Gemini was a large one, their relative similarity can be best underlined by the fact that only two unmanned tests were required prior to manned flight in Gemini. For Apollo, however, no fewer than twenty-two unmanned tests were conducted. This was two more than NASA required before the manned Project Mercury flights.

MISSION*	DESIG-NATION	DATE	TIME	SPACE-CRAFT*	LAUNCH VEHICLE	RECOVERY AREA	RECOVERY SHIP	OBJECTIVE*	RE-SULT*
SA-1	—	10/27/60	13:00:06	—	Saturn 1	—	—	Live first stage; dummy upper stages; first Saturn	S
SA-2	—	4/25/62	09:00:34	—	Saturn 1	—	—	Project High Water I	S
SA-3	—	11/16/62	12:45:02	—	Saturn 1	—	—	Project High Water II	S
SA-4	—	3/28/63	15:11:55	—	Saturn 1	—	—	Engine shutdown	S
PA-1	—	11/7/63	09:00:01	BP-6	—	—	—	Simulated pad abort	S
SA-5	—	1/29/64	11:25:01	—	Saturn 1	—	—	Orbital flight test	S
Abort test	A-001	5/13/64	05:59:59	BP-12	Little Joe 2	—	—	Verify dynamic shape of LES	S
SA-6	A-101	5/28/64	13:07:00	BP-13	Saturn 1	—	—	Orbital flight test of Saturn/Apollo	S
SA-7	A-102	9/18/64	11:22:43	BP-15	Saturn 1	—	—	Final qualification flight for Saturn 1	S
Abort test	A-002	12/8/64	08:00:00	BP-23	Little Joe 2	—	—	Guidance limits	S
SA-9	A-103	2/16/65	09:37:03	BP-16	Saturn 1	—	—	BP-16 contained Pegasus meteoroid satellite	S

Abort test	A-003	5/19/65	06:01:04	BP-22	Little Joe 2	—	—	High altitude abort test	F
SA-8	A-104	5/25/65	15:35:01	BP-26	Saturn 1	—	—	BP-26 contained Pegasus meteoroid satellite	S
PA-2	—	6/29/65	06:00:01	BP-23A		—	—	Simulated pad abort	S
SA-10	A-105	7/30/65	09:00:00	BP-9	Saturn 1	—	—	BP-9 contained Pegasus meteoroid satellite	S
Abort test	A-004	1/20/66	08:17:01	CSM-002	Little Joe 2	—	—	High altitude abort test	S
AS-201	—	2/26/66	11:12:01	CSM-009	Saturn 1B	Atlantic	USS *Boxer*	Launch vehicle/ spacecraft compatibility	S
AS-203	—	7/5/66	10:53:17	—	Saturn 1B	—	—	Orbital test of S-4B second stage	S
AS-202	—	8/25/66	13:55:32	CSM-011	Saturn 1B	Pacific	USS *Hornet*	Heat shield qualification	S
AS-501	Apollo 4	11/9/67	07:00:01	CSM-017 LTA-10R	Saturn 5	Pacific	USS *Bennington*	High apogee test of launch vehicle/ spacecraft	S
AS-204	Apollo 5	1/22/68	17:48:08	LM-1	Saturn 1B	—	—	Orbital test of first production lunar module	S
AS-502	Apollo 6	4/4/68	07:00:01	CM-020 SM-014 LTA-2R	Saturn 5	Pacific	USS *Okinawa*	High oxygen test of launch vehicle/ spacecraft	P

*SA: Saturn Apollo   PA: Pad Abort   AS: Apollo Saturn   LES: Launch Escape System   BP: Boilerplate
CSM: Command and Service Modules (CM, SM)   LTA: Lunar Test Article   LM: Lunar Module   S: Success   F: Failure

Project Apollo MANNED MISSIONS. This table provides comparative information on the eleven manned Apollo missions, six of which actually landed on the Moon. Crew assignments for primary and backup crews read in order: commander, command module pilot, lunar module pilot. Mission names read in order: command module, lunar module.

MISSION	ASTRONAUT	AGE	BACKUP	DATE	LAUNCH TIME	DURATION	RECOVERY	RECOVERY SHIP
Apollo 7	Schirra, Eisele, Cunningham	45 38 36	Stafford, Young, Cernan	10/11/68	11:02:45	260:09:03	Atlantic	USS *Essex*
Apollo 8	Borman, Lovell, Anders	40 40 35	Armstrong, Aldrin, Haise	12/21/68	07:51:00	147:00:02	Pacific	USS *Yorktown*
Apollo 9 *Gumdrop, Spider*	McDivitt, Scott, Schweickart	39 36 33	Conrad, Gordon, Bean	3/3/69	11:00:00	241:00:54	Atlantic	USS *Guadalcanal*
Apollo 10 *Charlie Brown, Snoopy*	Stafford, Young, Cernan	38 38 35	Cooper, Eisele, Mitchell	5/18/69	12:49:00	192:03:23	Pacific	USS *Princeton*
Apollo 11 *Columbia, Eagle*	Armstrong, Collins, Aldrin	38 38 39	Lovell, Anders, Haise	7/16/69	09:32:00	195:18:35	Pacific	USS *Hornet*
Apollo 12 *Yankee Clipper, Intrepid*	Conrad, Gordon, Bean	39 40 37	Scott, Worden, Irwin	11/14/69	11:22:00	244:36:25	Pacific	USS *Hornet*
Apollo 13 *Odyssey, Aquarius*	Lovell, Swigert,*** Haise	42 38 36	Young, ***, Duke	4/11/70	14:13:00	142:54:41	Pacific	USS *Iwo Jima*

*** Swigert became a member of the prime crew of Apollo 13 when he moved up from the backup crew to replace Mattingly who was exposed to measles prior to the launch.

Apollo 14 *Kitty Hawk, Antares*	Shepard, Roosa, Mitchell	47 37 40	Cernan, Evans, Engle	1/31/71	16:03:00	216:01:57	Pacific	USS *New Orleans*
Apollo 15 *Endeavour, Falcon*	Scott, Worden, Irwin	39 38 41	Gordon, Brand, Schmitt	7/26/71	09:34:00	295:11:53	Pacific	USS *Okinawa*
Apollo 16 *Casper, Orion*	Young, Mattingly, Duke	41 36 36	Haise, Roosa, Mitchell	4/16/72	12:54:00	265:51:05	Pacific	USS *Ticonderoga*
Apollo 17 *America, Challenger*	Cernan, Evans, Schmitt	38 39 37	Young, Roosa, Duke	12/7/72	00:33:00	301:51:59	Pacific	USS *Ticonderoga*

Skylab/Apollo-Soyuz MANNED MISSIONS. There were three missions to the orbiting Skylab space station in 1973–1974, plus one joint U.S.-Soviet mission called the Apollo-Soyuz Test Project. The Skylab missions lasted for 28, 59, and 84 days respectively, while the ASTP lasted 9 days.

MISSION	ASTRONAUTS	DATE	DURATION	RECOVERY SHIP
Skylab 2	Conrad, Kerwin, Weitz	5/25–6/22/73	672:49:49	USS *Ticonderoga*
Skylab 3	Bean, Garriott, Lousma	7/28–9/25/73	1,427:09:04	USS *New Orleans*
Skylab 4	Carr, Gibson, Pogue	11/16/73–2/8/74	2,017:15:32	USS *New Orleans*
Apollo-Soyuz Test Project	U.S.: Stafford, Brand, Slayton	7/15–7/24/75	217:28:24	USS *New Orleans*
	U.S.S.R.: Leonov, Kubasov		142:30:54	landed in U.S.S.R.

Space Shuttle Missions Approach and Landing Tests (ALT). A series of test flights was conducted with the orbiter *Enterprise* at NASA's Dryden Flight Research Center, Edwards Air Force Base, California, in 1977. For the "captive" tests, the *Enterprise* was bolted to the top of a modified Boeing 747 jet, where it remained for the entire flight. These tests measured aerodynamic properties and some onboard systems. On later missions, the *Enterprise* was dropped from the jet and piloted to a runway landing. These approach and landing tests were the first manned flights of the shuttle.

UNMANNED CAPTIVE FLIGHTS

MISSION	DATE	DURATION
1	2/18/77	2 hrs, 5 min
2	2/22/77	3 hrs, 13 min
3	2/25/77	2 hrs, 28 min
4	2/28/77	2 hrs, 11 min
5	3/2/77	1 hr, 39 min

MANNED CAPTIVE FLIGHTS

MISSION	ASTRONAUTS	DATE	DURATION
1	Haise, Fullerton	6/18/77	0:55:46
2	Engle, Truly	6/28/77	01:02:00
3	Haise, Fullerton	7/26/77	01:00:00

MANNED FREE FLIGHTS

MISSION	ASTRONAUTS	DATE	DURATION	COMMENTS
ALT 1	Haise, Fullerton	7/12/77	0:05:21	Tail cone on.
ALT 2	Engle, Truly	9/13/77	0:05:28	Tail cone on.
ALT 3	Haise, Fullerton	9/23/77	0:05:34	Tail cone on.
ALT 4	Engle, Truly	10/12/77	0:02:34	Tail cone off.
ALT 5	Haise, Fullerton	10/26/77	0:02:01	Tail cone off.

Orbital Flight Test Series (OFT)

MISSION	ASTRONAUTS*	BACKUP	LAUNCH DATE	DURATION	COMMENTS*
STS-1 (OFT-1) *Columbia*	Young (C), Crippen (P)	Engle, Truly	4/12/81	2.5 days	Two-day launch delay caused by computer problems. Landed on Runway 23, Edwards AFB.
STS-2 (OFT-2) *Columbia*	Engle (C), Truly (P)	Mattingly, Hartsfield	11/12/81	2.5 days	Five-day mission cut short after problems with onboard fuel cells. Landed on Runway 23, Edwards AFB.
STS-3 (OFT-3) *Columbia*	Lousma (C), Fullerton (P)	Mattingly, Hartsfield	3/22/82	8 days	Landed at White Sands, N. Mex., when bad weather drenched Edwards strips.
STS-4 (OFT-4) *Columbia*	Mattingly (C), Hartsfield (P)	Back-up crews no longer named.	6/27/82	7 days	Problems closing payload bay doors.

Operational Flights

MISSION	ASTRONAUTS*	LAUNCH DATE	DURATION	COMMENTS*
STS-5 *Columbia*	Brand (C), Overmyer (P), Allen (MS), Lenoir (MS)	11/11/82	5 days	Deployed 2 commercial communication satellites. First 4-man crew in space. First U.S. launch without pressure suits.
STS-6 *Challenger*	Weitz (C), Bobko (P), Peterson (MS), Musgrave (MS)	4/4/83	5 days	First shuttle EVAs, Musgrave and Peterson. Musgrave second U.S. M.D. in space.

* C: Commander, P: Pilot, MS: Mission Specialist, PS: Payload Specialist, KSC: Kennedy Space Center

Orbital Flight Test Series (OFT) *(continued)*

MISSION	ASTRONAUTS*	BACKUP	LAUNCH DATE	DURATION	COMMENTS*
STS-7 *Columbia*	Crippen (C), Hauck (P), Ride (MS), Fabian (MS), Thagard (MS)		6/18/83	6 days	First U.S. woman in space. First man to ride shuttle twice. Experiments on spacesickness. First landing attempt at KSC scrubbed for bad weather, landed Edwards.
STS-8 *Challenger*	Truly (C), Brandenstein (P), Bluford (MS), Gardner (MS), Thornton (MS)		9/30/83	6 days	First U.S. black in space. First night launch, first night landing.
STS-9 *Columbia*	Young (C), Shaw (P), Parker (MS), Garriott (MS), Lichtenberg (PS), Merbold (PS)		11/28/83	10 days	First non-American on U.S. mission (Merbold, West Germany). Edwards landing. First flight of ESA Spacelab. Sixth space mission for Young.
41-B *Challenger*	Brand (C), R. Gibson (P), McCandless (MS), McNair (MS), Stewart (MS)		2/3/84	7 days	McNair second U.S. black in space. First landing at KSC. First untethered EVA, using jet backpack (McCandless).
41-C *Challenger*	Crippen (C), Scobee (P), Nelson (MS), Hart (MS), van Hoften (MS)		4/6/84	7½ days	Crip's third shuttle flight. First attempt to recover satellite (Solar Maximum) with rocket backpack. Satellite recovered with remote manipulator, repaired, relaunched. Bad weather in Florida diverted landing to Edwards.

Mission / Orbiter	Crew	Date	Duration	Notes
41-D *Discovery*	Hartsfield (C), Coats (P), Mullane (MS), Resnick (MS), Hawley (MS), Charles Walker (PS)	8/30/84	7 days	Mission delayed after launch abort 6 seconds before lift off. Test of solar power generator. First commercial passenger (Walker) attempted to manufacture pharmaceutical products.
41-G *Challenger*	Crippen (C), McBride (P), Leetsma (MS), Ride (MS), Sullivan (MS), Scully-Power (PS), Marc Garneau (PS, Canada)	10/5/84	8 days	Test of satellite refuelling procedures; first U.S. woman to fly twice (Ride), first U.S. woman EVA (Sullivan). Largest crew to date. First Group 9 naut to fly (Leetsma), Canadian guest (Garneau). Extensive Earth mapping.
51-A *Discovery*	Hauck (C), D. Walker (P), J. Allen (MS), D. Gardner (MS), A. Fisher	11/8/84	8 days	Retrieval of two communications satellites for repair.

* C: Commander, P: Pilot, MS: Mission Specialist, PS: Payload Specialist, KSC: Kennedy Space Center

Soviet Manned Missions Vostok and Voskhod, 1961–65

MISSION	COSMONAUTS	AGE	BACKUP	LAUNCH DATE	DURATION	COMMENTS
Vostok 1 Cedar	Yuri Gagarin	27	Titov	4/12/61	1 hr 48 min	First man in space. First man in orbit.
Vostok 2 Eagle	Gherman Titov	27	Nikolayev	8/6/61	25 hours	First to eat and sleep in space. First overnight. First spacesickness.
Vostok 3 Falcon	Andrian Nikolayev	33	Bykovsky	8/11/62	4 days	First joint mission, came within 3 miles of rendezvous.
Vostok 4 Golden Eagle	Pavel Popovich	33	Komarov	8/12/62	3 days	
Vostok 5 Hawk	Valery Bykovsky	29	Volynov	6/14/63	5 days	Second joint mission. First woman in space. Vostok 6 duration longer than all Mercury flights combined.
Vostok 6 Seagull	Valentina Tereshkova	26	"Venera"*	6/16/63	3 days	
Voskhod 1 Ruby	Vladimir Komarov, Boris Yegorov, Konstantin Feoktistov	37 27 38	Volynov ? ?	10/12/64	1 day	First 3-man mission. First M.D. in space. First civilian in space.
Voskhod 2 Diamond	Pavel Belyayev, Alexei Leonov	39 31	? Khrunov	3/18/65	1 day	First spacewalk. First manual control of Soviet spacecraft. First Soviet manual re-entry and landing.

* Code name. Identities of other 5 women in program never revealed.

Soyuz Flights

MISSION	COSMONAUTS	AGE	BACKUP	LAUNCH DATE	DURATION	COMMENTS
Soyuz 1 *Ruby*	Vladimir Komarov	39	Gagarin	4/23/67	1 day	Craft tangled in its parachute lines during re-entry, pilot died on impact. First space death.
Soyuz 3 *Argon*	Georgi Beregovoy	47	Shatalov	10/26/68	4 days	First successful Soyuz flight.
Soyuz 4 *Amur*	Vladimir Shatalov	41	Filipchenko	1/14/69	3 days	First successful Soviet rendezvous and docking. First exchange of crew.
Soyuz 5 *Baikal*	Boris Volynov, Alexei Yeliseyev, Yevgeny Khrunov	34 34 35	Shonin, Kubasov, Gorbatko	1/15/69	3 days	
Soyuz 6 *Antaeus*	Georgi Shonin, Valery Kubasov	34 34	? ?	10/11/69	5 days	First 3-vessel joint mission. Docking attempt between Soyuz 7 and 8 failed.
Soyuz 7 *Snowstorm*	Anatoly Filipchenko, Vladislav Volkov	41 34	? ?	10/12/69	5 days	
Soyuz 8 *Granite*	Vladimir Shatalov, Alexei Yeliseyev	42 35	Nikolayev, Sevastyanov	10/13/69	5 days	
Soyuz 9 *Falcon*	Andrian Nikolayev, Vitaly Sevastyanov	40 35	Lazarev, Makarov	6/1/70	15 days	First Vostok veteran to fly again.
Soyuz 10 *Granite*	Vladimir Shatalov, Alexei Yeliseyev, Nikolai Rukavishnikov	44 37 39	? ? ?	4/22/71	2 days	Unsuccessful Salyut docking attempt.
Soyuz 11 *Amber*	Georgi Dobrovolsky, Vladislav Volkov, Viktor Patsayev	43 36 38	? ? ?	6/6/71	24 days	First successful docking with Salyut space station. All men died during re-entry after valve vented air out.

Soyuz Flights (*continued*)

MISSION	COSMONAUTS	AGE	BACKUP	LAUNCH DATE	DURATION	COMMENTS
Soyuz 12 *Urals*	Vasily Lazarev, Oleg Makarov	45 40	Gubarev, Grechko	9/27/73	2 days	Shakedown mission of new model.
Soyuz 13 *Caucasus*	Pyotr Klimuk, Valentin Lebedev	31 31	? ?	12/18/73	8 days	Astronomical research.
Soyuz 14 *Golden Eagle*	Pavel Popovich, Yuri Artyukhin	44 44	Volynov, Zholobov	7/3/74	16 days	Earth observation and military reconnaissance from Salyut.
Soyuz 15 *Danube*	Gennadi Sarafanov, Lev Dyomin	32 48	Zudov, Rozhdestvensky	8/26/74	2 days	Unsuccessful Salyut docking attempt.
Soyuz 16 *Snowstorm*	Anatoly Filipchenko, Nikolai Rukavishnikov	46 42	Romanenko, Ivanchenko	12/6/74	6 days	Simulation of Apollo-Soyuz Test Project.
Soyuz 17 *Zenith*	Alexei Gubarev, Georgei Grechko	44 44	? ?	1/10/75	30 days	Science experiments on Salyut.
Soyuz "X" *Urals*	Vasily Lazarev, Oleg Makarov	46 42	Klimuk, Sevastyanov	4/5/75	c. 15 min	Postlaunch abort, crash landing in Altai mountains; injuries.
Soyuz 18 *Caucasus*	Pyotr Klimuk, Vitaly Sevastyanov	33 40	Kovalyonok, Lebedev	5/24/75	63 days	Salyut mission overlapped Apollo-Soyuz flight.
Soyuz 19 *Soyuz*	Alexei Leonov, Valery Kubasov	41 40	Dzhanibekov, Andreyev	7/15/75	6 days	Apollo-Soyuz Test Project. First live TV coverage of Soviet launch.
Soyuz 21 *Baikal*	Boris Volynov, Vitaly Zholobov	43 39	? ?	7/6/76	49 days	Salyut mission.
Soyuz 22 *Hawk*	Valery Bykovsky, Vladimir Aksyonov	42 41	Malyshev, Strekalov	9/15/76	8 days	Earth mapping and reconnaissance. Nearly polar orbit.
Soyuz 23 *Rodon*	Vyacheslav Zudov, Valery Rozhdestvensky	34 37	Gorbatko, Glazkov	10/14/76	2 days	Salyut docking failed. First Soviet water landing, Lake Tengiz (accidental).

Craft	Crew	Age	Backup	Date	Duration	Notes
Soyuz 24 *Terek*	Viktor Gorbatko, Yuri Glazkov	43 38	? ?	2/7/77	18 days	Closeout for Salyut 5.
Soyuz 25 *Photon*	Vladimir Kovalyonok, Valery Ryumin	35 38	Romanenko, Ivanchenkov	10/9/77	2 days	Unsuccessful Salyut docking attempt.
Soyuz 26 *Tamyr*	Georgei Grechko, Yuri Romanenko	46 33	Ivanchenkov, Kovalyonok	12/10/77	96 days	First Salyut 6 mission. Near-fatality during spacewalk.
Soyuz 27 *Pamir*	Vladimir Dzhanibekov, Oleg Makarov	36 45	Kovalyonok, Ivanchenkov	1/10/78	6 days	First "trade-in" of Soyuz craft at Salyut.
Soyuz 28 *Zenith*	Alexei Gubarev, Vladimir Remek	47 29	Rukavishnikov, Oldrich Pelczak	3/2/78	8 days	First "guest cosmonaut," from Czechoslovakia.
Soyuz 29 *Photon*	Vladimir Kovalyonok, Alexander Ivanchenkov	36 38	Lyakhov, Ryumin	6/15/78	140 days	Salyut 6 mission.
Soyuz 30 *Caucasus*	Pyotr Klimuk, Miroslav Hermaszewski	36 37	Kubasov, Zenon Jankowski	6/27/78	8 days	Guest cosmonaut from Poland performed crystal growth experiments.
Soyuz 31 *Hawk*	Valery Bykovsky, Sigmund Jähn	44 41	Gorbatko, Eberhard Köllner	8/26/78	8 days	East German guest cosmonaut.
Soyuz 32 *Proton*	Vladimir Lyakhov, Valery Ryumin	38 40	Popov, Lebedev	2/25/79	175 days	Another endurance run aboard Salyut 6.
Soyuz 33 *Saturn*	Nikolai Rukavishnikov, Georgi Ivanov	47 40	Romanenko, Alexandrov	4/10/79	2 days	Salyut docking failed. Bulgarian guest cosmonaut.
Soyuz 35 *Dnieper*	Leonid Popov, Valery Ryumin	35 41	Kovalyonok, Strekalov	4/9/80	185 days	Endurance run aboard Salyut 6.
Soyuz 36 *Orion*	Valery Kubasov, Bertalan Farkas	45 34	Dzhanibekov, Bela Magyari	5/26/80	8 days	Hungarian guest. Soyuz trade.
Soyuz T-2 *Jupiter*	Yuri Malyshev, Vladimir Aksyonov	39 44	Kizim, Makarov	6/5/80	4 days	Shakedown flight of new model.

Soyuz Flights (*continued*)

MISSION	COSMONAUTS	AGE	BACKUP	LAUNCH DATE	DURATION	COMMENTS
Soyuz 37 Terek	Viktor Gorbatko, Pham Tuan	46 29	Bykovsky, Bui Thahn Liem	7/23/80	8 days	Vietnamese guest is first nonwhite, first Oriental in space.
Soyuz 38 Tamyr	Yuri Romanenko, Arnaldo Tamayo-Mendez	36 39	Khrunov, José Lopez-Falcon	9/18/80	8 days	Cuban guest is first black, first Hispanic in space.
Soyuz T-3	Oleg Makarov, Leonid Kizim, Gennadi Strekalov	47 ? ?	? ? ?	11/27/80	13 days	First 3-man mission since Soyuz 11 fatalities.
Soyuz T-4 Photon	Viktor Savinykh, Vladimir Kovalyonok	? 39	? ?	3/12/81	90 days	Salyut mission. Possible problems.
Soyuz 39	Vladimir Dzhanibekov, Jugderdemidin Gurrugcha	39 34	Lyakhov, Maidarjabin Ganzorig	3/22/81	8 days	Mongolian guest cosmonaut.
Soyuz 40	Leonid Popov, Dumitru Prunariu	? 29	? Dumitru Dediu	5/14/81	8 days	Romanian guest cosmonaut.
Soyuz T-5	Anatoly Berezovoy, Valentin Lebedev	? 40	? ?	5/12/82	211 days	First mission on Salyut 7.
Soyuz T-6	Vladimir Dzhanibekov, Alexander Ivanchenkov, Jean-Loup Chrétien	40 42 43	? ? Patrick Baudry	6/24/82	8 days	First Western European in space, French guest cosmonaut.
Soyuz T-7	Leonid Popov, Alexander Serebrov, Svetlana Savitskaya	37 38 34	? ? ?	8/19/82	8 days	Second woman in space, 19 years after Tereshkova (Vostok 6).

Flight	Crew			Date	Duration	Notes
Soyuz T-8	Vladimir Titov,	?	?	4/20/83	2 days	Docking failure with Salyut 7.
	Gennadi Strekalov,	?	?			
	Alexander Serebrov	39	?			
Soyuz T-9	Vladimir Lyakhov	42	?	6/27/83	150 days	Suggestions in British press that cosmonauts would be stranded. They weren't.
	Alexander Alexandrov	?	?			
Soyuz "T-X"	Vladimir Titov			9/27/83	-abort-	Booster explosion on pad, cosmonauts use escape tower, survive with injuries.
	Gennadi Strekalov					
Soyuz T-10	Leonid Kizim,	?	?	2/8/84	237 days	Endurance record through end of 1984.
	Vladimir Solovyev,	?	?			
	Oleg Atkov	?	?			
Soyuz T-11	Yuri Malyshev,	43	?	4/7/84	8 days	Guest cosmonaut from India tested effects of yoga to counter spacesickness. Live TV launch coverage.
	Gennadi Strekalov,	?	?			
	Rakesh Sharma	?	?			
Soyuz T-12	Vladimir Dzhanibekov	?	?	7/17/84	8 days	First woman to fly in space twice; first woman to walk in space.
	Svetlana Savitskaya,	?	?			
	Igor Volk	?	?			

Bibliography

Aldrin, Edwin E., Jr., *Return to Earth*, Random House, New York, 1973.

Armstrong, Neil; Edwin E. Aldrin, Jr.; Michael Collins, et al., *First on the Moon*, Little, Brown, Boston, 1970.

Asimov, Isaac, *Extraterrestrial Civilization*, Crown Publishers, New York, 1979.

—— *View from a Height*, Doubleday, New York, 1963.

Baker, David, *The History of Manned Space Flight*, Crown Publishers, New York, 1982.

Bergaust, Erik, *Murder on Pad 34*, Putnam, New York, 1968.

—— *Reaching For The Stars*, Doubleday, New York, 1960.

—— *Rocket City, USA*, Macmillan, New York, 1963.

Bova, Ben, *The High Road*, Houghton Mifflin, Boston, 1981.

Braun, Wernher von, *Space Frontier*, Holt, Rinehart & Winston, New York, 1963.

Caiden, Martin, *The Astronauts*, E. P. Dutton, New York, 1961.

Cipriano, Anthony, *America's Journey into Space*, Wanderer Books, New York, 1979.

Collins, Michael, *Carrying the Fire*, Farrar, Straus & Giroux, New York, 1974.

Cooper, Henry S. F., Jr., *Apollo on the Moon*, Dial Press, New York, 1969.

—— *13: The Flight that Failed*, Dial Press, New York, 1972.

Cunningham, Walt, *The All-American Boys*, Macmillan, New York, 1977.

Daniloff, Nicholas, *The Kremlin and the Cosmos*, Knopf, New York, 1972.

Dornberger, Walter, *V-2*, Viking Press, New York, 1954.

Fallaci, Oriana, *If the Sun Dies*, Atheneum, New York, 1966.

Franklin, Jon, and John Sutherland, *Guinea Pig Doctors*, William Morrow, New York, 1984.

Grissom, Betty, and Henry Still, *Starfall*, Crowell, New York, 1974.

Grissom, Virgil I., *Gemini: A Personal Account of Man's Venture into Space*, Macmillan, New York, 1968.

Ley, Willy, *Rockets, Missiles and Space Travel*, Viking, New York, 1961.

—— *Watchers of the Skies*, Viking, New York, 1963.

Mailer, Norman, *Of a Fire on the Moon*, Little, Brown, Boston, 1969.

McCullough, Joan, "Thirteen Who Were Left Behind," *Ms* magazine, September 1973.

Newton, Clarke, *1001 Questions Answered About Space*, Dodd, Mead, New York, 1971.

Oberg, James, *Red Star in Orbit,* Random House, New York, 1981.

O'Leary, Brian, *The Making of an Ex-Astronaut,* Houghton Mifflin, Boston, 1970.

Osman, Tony, *Space History,* St. Martin's Press, New York, 1983.

Peebles, Curtis, *The Battle for Space,* Beaufort Books, New York, 1983.

Roes, Nicholas, and William E. Kennedy, *Space Flight Encyclopedia,* Follett, New York, 1968.

Sagan, Carl, *Cosmos,* Random House, New York, 1980.

——— et al., *Murmurs of Earth,* Ballantine Books, New York, 1978.

Sharpe, Mitchell, *It Is I, Seagull,* Crowell, New York, 1975.

Sheldon, Charles, II, *Review of the Soviet Space Program with Comparative United States Data,* McGraw-Hill, New York, 1968.

Shepard, Alan B. Jr.; Virgil I. Grissom; John H. Glenn, Jr.; M. Scott Carpenter; Walter M. Schirra, Jr.; L. Gordon Cooper, Jr.; and Donald K. Slayton, *We Seven,* Simon & Schuster, New York, 1962.

Swenson, Loyd S. et al., *This New Ocean: A History of Project Mercury,* NASA Historical Series, Washington, D.C., 1966.

Sullivan, Walter, *We Are Not Alone,* McGraw-Hill, New York, 1964.

Wallechinsky, David; Irving Wallace, et al., *The People's Almanac* (2 vols.), Morrow, New York, 1975–76.

Wolfe, Tom, *The Right Stuff,* Farrar, Straus & Giroux, New York, 1979.

Glossary

abort Premature termination of an activity, as in *launch abort*. *See also* hold.

Agena An upper-stage rocket booster developed by Lockheed in 1959. It served as a booster for unmanned satellites and probes, and as a docking target vehicle in Project Gemini.

albedo Degree of light reflected back from an object in space. If it could reflect back all the light it receives, it would be albedo 1.0; the Moon is therefore a dull object at albedo 0.07.

ALFA Air lubricated free axis trainer; an early flight simulator.

AOS Acquisition of signal. The point in time or geography when radio communications can be reestablished with a spacecraft. During portions of each orbit, the craft is out of range of ground-based radio systems (loss of signal). When the craft comes back into range, "We have AOS."

apogee In an elliptical orbit, the point farthest from the Earth.

Apollo The NASA project that included the manned lunar landings, 1969–1972.

Apollo Applications Project (AAP) A project which planned to use data gathered during Project Apollo for long-term Earth-orbital flights. Later incorporated into Skylab.

Apollo-Soyuz Test Project (ASTP) The joint mission in 1975 of three American and two Soviet flyers. Called a test project since only one mission was planned.

AS Apollo-Saturn. An Apollo spacecraft atop a Saturn booster. AS-7, for example, was the seventh time an Apollo-Saturn combination had been launched.

Atlas A General Dynamics rocket originally designed as an ICBM, it served as the booster for four Mercury missions.

attitude Position of an object in relation with another. Flying the Shuttle nose-to-sun is a different attitude than flying nose-to-ground. Sometimes confused with altitude.

back out After an abort, the process of securing the facilities, usually by running through the countdown sequence in reverse. (Disarm the explosive bolts, disarm the rockets, unload the fuel, etc.)

ballistic trajectory A path in space like a bullet, or similar thrown object. Pre-Shuttle spacecraft had very little control over their own flight paths, and so depended heavily on proper alignment at launch and at orbital insertion.

BECO   Booster engine cut-off. The point after launch when the engines are shut down and the spacecraft begins to coast. *See also* MECO and VECO.

bird   Aircraft or spacecraft. "The pilot ejected, but we lost the bird."

blast off   Archaic version of *lift off*. Used only by pulp science fiction writers and geriatric journalists.

boilerplate   A heavy test copy of a spacecraft, used for ground tests but not for actual flight. Also, a derisive term for Soviet spacecraft because of their comparative size and weight.

burn   A brief firing of onboard engines, usually to correct attitude or course.

CapCom   Contracted from "capsule communicator," the person (always an astronaut) who is in voice communication with the flight crew. With the exception of a few senior NASA officials, all communication with astronauts in flight must be channelled through the CapCom.

Cape Canaveral   (1) Geographical feature on Florida's east-central coast. (2) Cape Canaveral Air Force Station, site of most U.S. unmanned space launches, and the manned launches in the Mercury and Gemini projects. (3) City directly south of that Air Force Station. (4) By common error, the Kennedy Space Center on Merritt Island, site of the Apollo, Skylab, and Space Shuttle launches.

Cape Kennedy   The name given to both the city of Cape Canaveral and the adjacent Air Force Station from 1963 to 1973, and by common error, to the Kennedy Space Center.

capsule   Archaic term for spacecraft. Although NASA stopped using it during Project Mercury, some reporters continued using the term during Shuttle flights.

centrifuge   A mechanism for inducing centrifugal force on a person or object. Often built as a container on a long arm revolving around a central point.

cislunar   The space between the surface of the Earth and the orbit of the Moon.

CM   Command module. The cone-shaped tip of the Apollo spacecraft, and the only portion to return to Earth.

comsat   Communications satellite. Also, Comsat General Corporation.

countdown   Schedule of events leading to launch, noted in a reverse time line.

cryogenic   Class of fuels that are liquid only at very low temperatures. Oxygen liquifies at $-297°F$, hydrogen at $-422°F$.

cryoloading   Loading of the cryogenic fuels into onboard fuel tanks, usually just a few hours before launch.

CSM   Command and service module. The CM and the cylindrical segment just below it, where Apollo astronauts did much of their work.

delta   In engineering, a change or difference between two numbers, such as a change in velocity. Used commonly to mean an unexpected problem, or departure from the original plan.

Delta   Booster rocket built by McDonnell-Douglas and used primarily for putting small to medium sized satellites into Earth orbit.

destruct system   Radio-activated explosive charges aboard a rocket booster. If the rocket goes too far off course during launch, the range safety officer sends the destruct signal, and the spacecraft comes down in little pieces, preferably away from populated areas. Also known as "hitting the *panic button*."

direct ascent   Launching directly from Earth to the target in space, without going first into a parking orbit around the Earth. Done only with lighter lunar probes.

docking   Connecting two spacecrafts while in orbit. *Also called* link-up.

Dyna-Soar   Contraction of dynamic soaring. A spaceplane that would have been launched vertically atop a Titan rocket, then would land on an airstrip. Project cancelled in 1963 in favor of Apollo; later reincarnated as the Space Shuttle. Also called X-20.

ERTS   Earth Resources Observation Satellite. An orbital observation and mapping spacecraft, renamed Landsat.

ESA   European Space Agency. A consortium of aerospace corporations spanning most of western Europe, developing space hardware and eventually their own launch systems. Most visible success to date has been the joint Spacelab project with NASA.

EVA   Extra-vehicular activity. Activity outside a spacecraft; a *spacewalk* or *moonwalk*.

FIDO   Flight dynamics officer, responsible for propulsion systems during a mission.

FLATs   First lady astronaut trainees. Group of thirteen women trained for possible spaceflight between 1959 and 1962.

footprint   The geographic area where debris is likely to fall when a spacecraft re-enters the atmosphere, breaks up, and crashes to Earth.

fuel cell   A device aboard a spacecraft which combines hydrogen and oxygen to provide water and electricity.

Gemini   NASA's second manned space project, using a two-man spacecraft, 1965–1966. Established ability of astronauts to function outside of spacecraft, to rendezvous and dock with other spacecraft.

g-force   Gravitational force; one g is the amount of force needed to move an object at 32.2 feet per second, equal to the Earth's gravitational attraction at sea level. An astronaut who is rapidly accelerating or decelerating goes through additional g-stress, or feels like he or she is on a planet with greater gravity.

GIRD   Russian abbreviation for "Group for the Investigation of Rocket Dynamics," which launched the first Soviet test rocket in the 1930s, under the direction of Tsander and Korolyov. Also translated as Group for the Study of Propulsion Dynamics.

GT   Gemini-Titan.   Gemini spacecraft mounted to a Titan booster.

GUIDO   Guidance officer, responsible for navigation and guidance during a mission.

hold   A delay, usually in the countdown sequence. A built-in hold is a safety margin written into the schedule to allow technicians to catch up on any work left undone to that point. A number of ground workers have the authority to *call a hold* if they detect a problem.

hypergolic   Fuels that ignite on contact with each other. Nitrogen tetroxide and mono-methyl-hydrazine are *hypergolic* with each other. These substances, used in auxillary power units and small maneuvering engines, tend to be highly toxic and corrosive.

ICBM Inter-continental ballistic missile. Rocket designed to carry a nuclear warhead at least 3,000 miles. Both U.S. and Soviet space programs began by using ICBMs to carry space travelers.

ignition Turning on an engine.

injection A burn to take a spacecraft out of orbit, toward another destination. *Trans-Lunar Injection* is the burn that took the Apollo spacecraft from Earth orbit to the Moon.

insertion Placement of a spacecraft into a preliminary, highly elliptical orbit. Also called *orbital insertion*.

JATO Jet assisted take-off.

JPL Jet Propulsion Laboratory, Pasadena, CA. A NASA research center operated in conjunction with California Institute of Technology. Best known for planetary research. *See* Viking and Voyager.

JSC Johnson Space Center. NASA Mission Control Center at Clear Lake, Texas, near Houston. Formerly called Manned Spacecraft Center, or simply "Manned Space."

KSC Kennedy Space Center. A massive complex on Merritt Island, FL, adjacent to Cape Canaveral. Site of the Apollo, Skylab, and Space Shuttle launches.

LANDSAT A series of U.S. observation satellites used for mapping, geology, and determining locations of resources both on land and underwater.

launch window The period of time within which launch must occur if mission objectives are to be met. If the window is missed, there may be another window the following day. On interplanetary missions, the windows tend to be the smallest, sometimes only a few minutes in duration.

lift off The spacecraft's ascent to at least two inches above the pad.

link-up *See* docking.

LM Lunar module, the portion of the Apollo spacecraft that actually landed on the Moon. In some early documents, it was called the *LEM,* or lunar excursion module, but that name was dropped as too cumbersome. Even so, since both acronyms were pronounced "lem," the earlier version turned up in many print accounts.

LOX Liquid oxygen.

Lunar rover A lightweight electric car taken to the Moon on later Apollo missions. It let astronauts move farther from the LM than would have been feasible by walking in cumbersome spacesuits.

MA Mercury-Atlas.

Mariner A series of unmanned probes that explored Mars, Venus and Mercury from the mid-1960s to mid-1970s.

MASTIF Multiple Axis Space Test Inertia Facility. Another early NASA "human stress evaluator," designed to induce dizziness or loss of direction.

MECO Main engine cut-off.

MER Manned earth reconaissance. U.S. Navy's late-fifties spaceflight proposal.

MILF Merritt Island Launch Facility. Old name for Kennedy Space Center.

MOCR Mission Operations Control Room. The nerve center of Mission Control.

MOL Manned orbiting laboratory. An Air Force project to use teams of astronauts to do scientific and medical research in a long-duration Earth-orbit space station. Cancelled in 1967. Many of the ideas became part of the Skylab experiments.

MR Mercury-Redstone.

MSC Manned Spacecraft Center. Gemini-Apollo era name for Johnson Space Center, Houston.

MSFC Marshall Space Flight Center. NASA facility in the Army's Redstone Arsenal, Huntsville, AL, where Dr. Wernher von Braun's team developed the Saturn rocket.

NACA National Advisory Committee on Aeronautics. Federal agency that did basic research into aviation from the 1920s through 1950s.

NASA National Aeronautics and Space Administration. Successor to NACA, established 1958, researching both aviation and the "peaceful exploration of space," according to the 1957 Space Act.

orbit The pathway of a spacecraft or celestial body in its revolution around another body, or a 360° circuit in space. By contrast, a revolution occurs when a spacecraft passes over a ground point after circling the body. Orbit is a space-based reference, revolution is ground-based. Above a stationary body which doesn't rotate, they would be the same, but when the Shuttle circles the Earth, the revolution takes about six minutes longer than the orbit.

passive communications satellite A satellite that simply reflects a radio signal without actively retransmitting or amplifying it. The Echo series.

payload The instrument package or manned spacecraft that a booster rocket carries into space. The freight as opposed to the train.

perigee Point of an orbit closest to the Earth.

Pioneer A series of interplanetary probes launched from 1958–1978. Pioneers 10 and 11 were the first to pass the orbit of Pluto; Pioneer Venus 1 and 2 discovered more about that planet than all previous research combined.

polar orbit An orbit that carries the spacecraft over both the north and south poles.

pressure suit Spacesuit.

Ranger A series of lunar probes that transmitted photos of potential landing sites back to Earth before crashing into the Moon. Early 1960s.

real-time Time as it is usually observed, as opposed to a delay. A real-time navigational system tells you where you are this second: a non-real-time system tells you where you were several minutes or hours ago, or where you will be at a given moment in the future.

Redstone A small (by modern standards) rocket booster used on the MR-3 and MR-4 flights of Shepard and Grissom. Named after the Army's Redstone Arsenal in Huntsville, Alabama.

re-entry Return of a spacecraft to the Earth's atmosphere. A crucial time, since atmospheric friction against the spacecraft hull can raise temperatures to almost 3,000°F.

rendezvous Maneuver bringing two or more spacecraft almost close enough to touch.

RETRO Retro-fire officer, responsible for calculations to fire retro rockets before re-entry. Along with FIDO and GUIDO, among the highest-ranking flight officers at Mission Control.

retro-fire Firing of rockets opposite to the direction of travel to slow down a spacecraft, as in re-entry.

roll-out The transfer of a fully built spacecraft from a hangar or VAB to the pad. The Shuttle is rolled out to the pad on what looks like a giant military tank.

Salyut   A series of seven (to date) Soviet space stations, which have had up to five occupants at once. Introduced in 1971, the tenth year of Soviet manned spaceflight, it was named as a "salute" to the late Yuri Gagarin.

Saturn   A series of rocket boosters that powered the Apollo and Skylab missions. The largest of these, the Saturn 5, generated 7.5 million pounds of thrust at lift-off, making it the most powerful rocket ever (as of 1984).

Skylab   U.S. space station which was manned by three teams of astronauts in 1973 and 1974. Crashed to Earth in 1979.

Soyuz   The Soviet three-man spacecraft. Since its introduction in 1967, the Soyuz has become the most-flown, most successful spacecraft in history. Soyuz, "union" in Russian, was the first Soviet spacecraft capable of docking.

Space Shuttle   Fleet of five spacecraft designed to take off vertically from a launch pad, orbit the Earth, and land horizontally on a landing strip. The Shuttle program was approved in 1972, with the first manned launch on April 12, 1981. *See* STS.

space station   Any large manned structure capable of docking with other spacecraft and serving as a base of operations.

SRB   Solid rocket booster. A rocket engine using solid rather than liquid fuel. The Space Shuttle uses a pair of giant SRBs attached to the sides of the liquid fuel tank for extra thrust at launch. Other rocket systems have used smaller SRBs for high initial thrust. Unlike liquid rockets, SRBs are essentially slow firecrackers and cannot be turned off once lit.

staging   The separation of an expended booster stage.

static firing   A test firing of rocket engines in which neither the engine nor spacecraft actually moves. Similar to firing a gun from a vise.

STS   Space Transportation System. Official designation of the Space Shuttle.

Titan   A series of large launch boosters, including the ones that carried the Gemini spacecraft on orbital missions. Built by Martin-Marietta.

TLI   Trans-lunar injection. Burn of rockets required to take a spacecraft out of Earth orbit and set it on course for lunar rendezvous.

Tyuratam   Site of the Soviet manned launch facility in Kazakhstan. Referred to in Soviet literature as the Baikonur Kosmodrom, or Star City.

VAB   Vehicle Assembly Building. Immense cube-shaped construction hangar at Kennedy Space Center where final assembly is completed on the Space Shuttle before roll-out to the pad. Once called Vertical Assembly Building.

VECO   Vernier engine cut-off.

vernier   Small auxilliary rocket used for guidance and added power during launch.

Viking   Two automated research stations that landed on Mars in 1976, transmitting data back to Earth for several years.

Voskhod   "Ascent." Soviet designation for a three-man mission and a two-man mission using modified Vostok spacecraft.

Vostok   "East." The Soviet Union's first manned spacecraft, used in six missions between 1961 and 1963.

Voyager   A pair of Mariner-type space probes launched in 1977 to explore the outer solar system. Transmitted magnificent photos of Jupiter in 1979, and of Saturn in 1981. Should encounter Uranus in 1986.

zero-g   Weightlessness. More correctly in Earth orbit, micro-g, since gravity is still quite detectable at 200 miles altitude.

Index